ANALYSIS OF
QUEUEING NETWORKS
WITH BLOCKING

INTERNATIONAL SERIES IN
OPERATIONS RESEARCH & MANAGEMENT SCIENCE

Frederick S. Hillier, Series Editor
Stanford University

ANALYSIS OF
QUEUEING NETWORKS
WITH BLOCKING

by

Simonetta Balsamo
Universita' di Venezia

Vittoria de Nitto Personé
Universita' di Roma "Tor Vergata"

Raif Onvural
IBM

KLUWER ACADEMIC PUBLISHERS
Boston / Dordrecht / London

Distributors for North, Central and South America:
Kluwer Academic Publishers
101 Philip Drive
Assinippi Park
Norwell, Massachusetts 02061 USA
Telephone (781) 871-6600
Fax (781) 871-6528
E-Mail <kluwer@wkap.com>

Distributors for all other countries:
Kluwer Academic Publishers Group
Distribution Centre
Post Office Box 322
3300 AH Dordrecht, THE NETHERLANDS
Telephone 31 78 6392 392
Fax 31 78 6546 474
E-Mail <orderdept@wkap.nl>

 Electronic Services <http://www.wkap.nl>

ISBN 978-1-4419-5014-7

Library of Congress Cataloging-in-Publication Data

A C.I.P. Catalogue record for this book is available from
the Library of Congress.

Printed on acid-free paper.

Printed in the United States of America

CONTENTS

5. Exact analysis of special networks

6. Approximate and bound analysis

PART III PROPERTIES OF NETWORKS WITH BLOCKING

7. Equivalence, insensitivity and monotonicity properties

PREFACE

Queueing network models have been widely applied as a powerful tool for modelling, performance evaluation and prediction of discrete flow system, such as computer systems, communication networks, production lines and manufacturing systems. Queueing network models with finite capacity queues and blocking have been introduced and applied as more realistic models of systems with finite capacity resources and with population constraints. This book provides an introduction to queueing networks with blocking and their properties and analysis.

In recent years, research in this field has progressed rapidly and many results have been presented in the literature. This book introduces queueing network models with finite capacity and various types of blocking mechanisms. It gives a comprehensive definition of the analytical model underlying these blocking queueing networks. It surveys exact and approximate analytical solution methods and algorithms and their relevant properties. It also presents various application examples of queueing networks to model computer systems and communication networks.

The book is organized in three parts. Queueing networks with blocking are introduced in part I, their analysis is presented in part II and various properties are defined in part III.

Specifically, part I introduces queueing networks with blocking and various application examples. The first chapter presents queueing networks, the main results and solution techniques to derive performance metrics, including the well-known class of product form BCMP networks.

Chapter 2 defines queueing networks with blocking, introduces the various blocking mechanisms and the specific definition of the performance indices. We discuss the relation between models with blocking and state dependent routing.

Chapter 3 presents application examples of queueing networks with blocking for modelling and performance evaluation of computer, communication and manufacturing systems.

Part II deals with exact and approximate analysis of queueing networks with blocking and the condition under which the various techniques can be applied.

Chapter 4 defines the exact analysis of queueing networks with blocking based on the definition and solution of a stochastic process associated to the queueing network. For each type of blocking model we present the complete definition of the continuous-time Markov chain underlying the network model. This allows the reader a complete definition of the exact solution algorithm for this class of models.

Chapter 5 presents some special cases. They include product form queueing networks with blocking and symmetrical networks. We present the models, the solution methods and the derivation of performance indices.

Chapter 6 reviews and compares approximate and bound algorithms and provides some guidelines in the selection of the approximate approach for each type of model.

Part III presents a review of various properties of networks with blocking. Chapter 7 describes several equivalence properties both between networks with and without blocking and between different blocking types. It presents equivalence properties in terms of average performance indices and passage time distribution, as well as the special case of fork/join networks. Other relevant properties are insensitivity and monotonicity of performance indices with respect to various network parameters.

Chapter 8 deals with the buffer allocation problem in queueing networks with finite capacity queues and presents approximate solution methods.

This book is partially based on the research of the authors. We would like to thank several colleagues and students for stimulating and helpful discussions. We would like to thank Lorenzo Donatiello for encouraging us to start this project. Thanks for discussions and their contributions to Ivan Mura, Claudia Clò, Francesco Ilario, Antonio Rainero. Raif Onvural especially thanks Professor Harry Perros for his mentorship during his studies and his friendship since then.

We would like to thank Frederick S. Hillier and Gary Folven for their constant encouragement and patience during the completion of this book.

PART I

QUEUEING NETWORK MODELS WITH BLOCKING AND APPLICATIONS

1 INTRODUCTION

System performance has been a major issue in the design and implementation of computer systems, production systems, communication systems, and flexible manufacturing systems. The success or failure of the design and operation of such systems is judged by the degree to which performance objective are met. Thus, tool and techniques for predicting their performance measures have received great attention in the research and development communities since early 1900s.

Queueing theory was developed to understand and to predict the behavior of real life systems. Conceptually, the simplest queueing model is the single queueing system illustrated in figure 1.1. The system models the flow of customers as they arrive, wait in the queue if the server is busy serving another customer, receive service, and eventually leave the system.

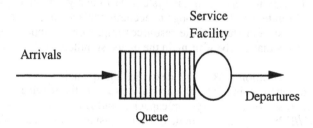

Figure 1.1. A single queueing system

To describe the behavior of a queueing system in time, five basic characteristics of the process need to be specified:

    i) the arrival pattern,
    ii) the number of servers,
    iii) the service pattern,
    iv) the service discipline,
    v) the queue capacity.

The arrival pattern or input to a queueing system is often measured in terms of the average number of arrivals per some unit of time, called the mean arrival rate. The reciprocal of the mean arrival rate is referred to as the average interarrival time. If the arrival pattern is deterministic, then the arrival process is fully characterized by the mean arrival rate. If the arrival pattern is random, further characterization is required in the form of the probability distribution associated with the process.

The number of service channels refers to the number of parallel servers that can service customers simultaneously. Each service channel may correspond to a physically or logically separate service facility with a common queue shared by all customers.

The rate at which customers are served is called the mean service rate. The reciprocal of the mean service rate is referred to as the average service time at the service facility. In the case of deterministic service processes, specification of the mean service rate is sufficient to describe the process, whereas if the process is random its probability distribution needs to be specified.

If a customer arrives at the system at a time the server(s) is unavailable to provide service to it (i.e., busy servicing another customer, failed and unavailable to provide service), it is forced to wait in the queue temporarily until it can start receiving service. If there are more than one customer waiting in the queue at a time the server becomes available (i.e., completion of a service or the service facility is repaired), one of the customers in the queue is selected to start receiving service. The manner by which customers are selected for service when a queue has formed is referred to as the service discipline.

Finally, the system capacity is the upper limit on the number of customers (waiting for and receiving service) in the system. Most analytical studies require the queue size to be infinite, i.e., large enough to accommodate all arriving customers. In practice, however, systems have finite resources imposing an upper bound on the number of customers that can be waiting in the queue simultaneously.

The Kendall's notation $A/B/X/Y/Z$ describes the queueing process of a single queueing system where A indicates the arrival pattern, B the service pattern, X the number of parallel servers, Y the system capacity, and Z the service discipline. For example, D/D/1//FCFS describes a single queueing system with deterministic arrival and service processes, one server, infinite system capacity (that is, there is always a space in the queue for arriving customers), and first come first served service discipline.

1.1   COMMONLY USED PROBABILITY DISTRIBUTIONS

The most common queueing models assume that the interarrival and service times are exponentially distributed. Equivalently, the arrival and service processes follow a Poisson distribution. That is, if the interarrival times are exponentially distributed then the number of arrivals at the system follows a Poisson distribution. Similarly, for the service process.

Consider an arrival process $\{N(t), t>0\}$ that satisfies the following assumptions, where $N(t)$ denotes the total number of arrivals up to time t, with $N(0)=0$:

i)   The probability that an arrival occurs between time t and t+Δt is equal to $\lambda\Delta t+o(\Delta t)$ where λ is a constant, Δt is an incremental element, and $o(\Delta t)$ denotes a quantity that becomes negligible as Δt goes to zero.

ii) The probability that there is more than one arrival between t and $\Delta t+t$ is $o(\Delta t)$.

iii) The numbers of arrivals in non-overlapping intervals are statistically independent.

Let $p_n(t)$ be the probability of n arrivals in a time interval t. Under the three assumptions above, we have:

$$p_n(t)=(\lambda t)^n\, e^{-t}/n! \qquad (1.1)$$

This distribution is referred to as the Poisson distribution with rate λ. As discussed above, if N(t) is Poisson with rate λ then the time between arrivals is exponentially distributed with mean $1/\lambda$. That is, let T be the random variable denoting time between two events (i.e., two arrivals or two service completions at a server), then:

$$\Pr\{T\leq t\}=1 - e^{-\lambda t} \qquad (1.2)$$

One of the interesting properties of the exponential distribution is the Markovian (also called memoryless) property, which states that the probability that a customer currently in service is completed at some future time t is independent of how long the customer has already been in service. It is mainly this special property that the exponential distribution has been the most widely used distribution in the analysis of queueing networks.

The variance of a distribution, var(X), gives a rough measure of spread. For example, let X denote the service time. If var(X) is small, the service times of customers are close to the mean with a little variation. Var(X) is equal to zero if X is deterministic (i.e., there is no variation). If a random variable is exponentially distributed with parameter λ, its variance is equal to the square of the mean service time (i.e., $1/\lambda^2$). Conversely, a large variance indicates that the service times of customers are widely spread and there is a large variation from the mean.

Finally, the squared coefficient of variation of the service time, c^2, gives a rough measure of the spread normalized over the square of the mean service time:

$$c^2= var(X)/\{mean(X)\}^2 \qquad (1.3)$$

The c^2 of a deterministic distribution is equal to zero whereas c^2 of exponentially distributed random variables is equal to one. In practice, the arrival and/or service time distributions are not known a priori and they often are not exponential. However, there are also many arrival and/or service time distributions that fit an exponential. The popularity of the exponential distribution arise out of the fact that it often yields to computationally efficient procedures and obtaining system behavior under this assumption is, in most cases, a relatively easy task. However, it is necessary to develop a framework that addresses more general distributions while keeping in mind that the model needs to be solved efficiently in order to obtain the system performance of interest.

Towards this goal, several probability distributions that are general and flexible enough to allow the relaxation of the assumption that the service and the interarrival times are exponentially distributed was defined. One approach is to use a simple exponential network to represent the service time required from a single server, as illustrated in figure 1.2.

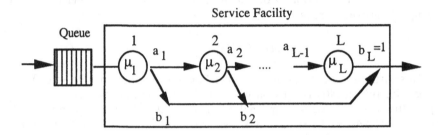

Figure 1.2. A queueing system with a Coxian server with L stages

Consider the service facility of this single queueing system. In this model, c^2 of the service distribution can be any value greater than zero with an appropriate choice of the parameters a_i, b_i, and L.

There can be at most one customer in nodes 1 to L at any time. Customers enter the service via node 1. The service time at node m is exponentially distributed with mean $1/\mu_m$. A customer completing its service at node m leaves the system with probability b_m or proceeds to node m+1 with probability a_m, ($a_m+b_m=1$, m=1,...,L-1). After node L, the customer leaves the system with probability one. This service distribution is referred to as a Coxian distribution with L stages. Any probability distribution function can be approximated arbitrarily closely by a Coxian distribution. This framework allows an arbitrary distribution that does not have the Markovian property to be approximated by a Coxian distribution that has the Markovian property.

For example, let us consider the service process of figure 1.2 with $b_i=0$, i=1,...,L-1 and assume that $\mu_i=\mu$, i=1,...,L. This is referred to as the Erlang distribution with L stages. The c^2 of this distribution is equal to $1/\sqrt{\lambda}$. As L goes to infinity, c^2 goes to zero approximating a deterministic distribution.

1.2   ANALYSIS OF A SINGLE QUEUE

A queueing process with Poisson arrivals, exponentially distributed service times, one server, with capacity B, and First Come First Served service discipline is referred to as an M/M/1/B/FCFS queue, where M stands for Markovian (memoryless). Let $\pi_n(t)$, $n \in E$, be the probability that there are n customers at time t in an M/M/1/B/FCFS queue, where n and E are respectively referred to as the *state* and the *state space* of the queueing system. For this queueing system, we have E={0,1,...,B}. Furthermore, let us assume that the system state (n) eventually becomes independent of the initial state so that no matter what time we query the system, the probability of finding n customers in the system remains constant, independent of time t. Then, π_n is referred to as the steady state probability of having n customers in the system.

The transition rate diagram of a process is a graphical representation of the transitions between the states of the process. The directed arcs between the states denote the one-step transitions of going from one state to the other. For example, transitions out of state n of an M/M/1/B/FCFS queue occurs either to state n+1 with an arrival (with rate λ) or to state n-1 with a departure (with rate μ). Its transition rate diagram is given in figure 1.3:

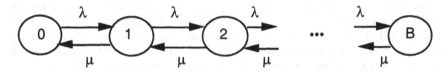

Figure 1.3. Transition rate Diagram of an M/M/1/B/FCFS queue

The global balance equations equate the total rate out of state n to the total rate into state n, for each state $n \in E$. The global balance equations of the M/M/1/B/FCFS queue can be easily written from the transition rate diagram given in figure 1.3:

$$\lambda \pi_0 = \mu \pi_1$$
$$(\lambda+\mu) \pi_n = \lambda \pi_{n-1} + \mu \pi_{n+1}, \; n=1,...,B-1$$
$$\lambda \pi_B = \mu \pi_{B-1}$$

In this system, there are B+1 unknowns (π_n's) and B+1 equations of which only B of them are linearly independent. As π_n is a probability distribution, the sum of all π_n's over its state space should add up to one, i.e. $\sum_{0 \leq n \leq B} \pi_n = 1$. This equation is referred to as the normalization equation. Using any B of the above B+1 equations together with the normalization equation, there are B+1 linearly independent equations, which can be solved to obtain π_n's. Writing these equations in matrix form, we have:

$$Q\pi = b$$

where
 b=(0,...,0,1) is a B+1 vector with 0's in positions 1 to B and 1 at position B+1;
 $\pi = (\pi_0,..., \pi_B)$ is a B+1 vector of steady state queue length probabilities;

$$Q = \begin{bmatrix} -\lambda & \mu & & & & \\ \lambda & -(\lambda+\mu) & \mu & & & \\ & \lambda & -(\lambda+\mu) & \mu & & \\ & & & \cdots & \cdots & \cdots \\ 1 & 1 & 1 & \cdots & 1 & 1 \end{bmatrix}$$

is a (B+1)x(B+1) matrix referred to as the rate matrix of the process.

In this rate matrix, we use the first B of the above B+1 equations and replaced the B+1st equation with the normalization equation.

1.3   QUEUEING NETWORKS

Customers in practice usually require different services provided by different servers. During this process, customers may have to wait in several different queues before receiving the required services. Such complex service systems can be modeled by defining a network of single queueing systems, referred to as a *queueing network*. A queueing network can be thought of as a connected directed graph whose nodes represent the service centers. Each node has a queue associated with it. The arcs between those nodes indicate the one-step moves that customers may make from one service center to another service center. The set of nodes and the set of arcs that connect the nodes are referred to as the *topology* of the network. The route that a customer takes through the network may be deterministic or random. Customers may be of different types and may follow different routes through the network. To complete the definition of a queueing network, the assumptions on the parameters of each node must be specified (i.e. the number of servers, the service discipline, the node capacity, and the service time distribution).

Queueing networks can be classified as open, closed and mixed.

In an open network model, customers enter the network from outside, receive service at one or more nodes according to the network topology, and eventually leave the network. Figure 1.4 illustrates an open network with five nodes, two

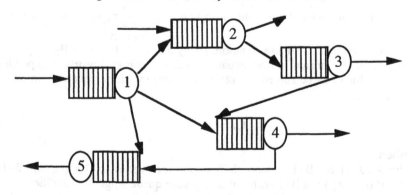

Figure 1.4. An Open Network

incoming arcs that denote arrival streams at nodes 1 and 2 and four outgoing arcs that represent customers' departure from the network.

We assume that the arrivals to an open network occur from an infinite external population of customers.

In closed queueing networks, no arrivals to or departures from the network are allowed and there is a constant population of customers circulating in the network. Figure 1.5 illustrates an example of a simple closed network with four nodes and N customers circulating in it. Since node 4 has N servers, then each customer arriving at that node finds a free server and it never waits in the queue. This type of node is called with infinite servers discipline. Such a closed network model can be used to represent a time sharing system as we shall discuss in chapter 3.

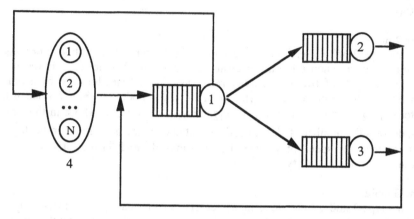

Figure 1.5. A A Closed Network

Mixed queueing networks are models with multiple types of customers that are closed networks with respect to some customer types and open with respect to others.

1.3.1 Product form solutions and BCMP theorem

Queueing networks have been studied in the literature under a multiplicity of assumptions. After some pioneering works, the development and analysis of one of the most general queueing networks is known as the BCMP theorem, bearing the initials of the authors Baskett, Chandy, Muntz and Palacios (1975). Although this theorem is applicable to both open and closed networks, we present the result only in the context of closed queueing networks.

Consider a closed queueing network with M nodes. Each node has an infinite capacity, that is there is always a space at a node for arriving customers. The topology of the network is arbitrary. There are R different types of customer classes. For the sake of simplicity here we assume that classes and chains have the same meaning. A customer of class r, upon completing service at node i goes to node j with probability $p_{ij;r}$. N_r is the number of class r customers circulating in the network. Let S and S_i denote respectively the state of the network and the state of node i, i.e. $S=(S_1,S_2,...,S_M)$ whereas S_i depends on the type of node i. In particular, node i can be one of the following four types, referred to as the BCMP nodes.

Type 1 node

All customers have the same service time distribution at a type 1 node which is exponentially distributed with mean $1/\mu_i$. Customers are served in the order of arrival, referred to as the first come first served (FCFS) service discipline. The state S_i of the node is defined as the vector $(r_1,r_2,...,r_{n_i})$, where n_i is the number of customers present at node i and r_j is the class index of the j-th customer in the FCFS order. There is a single server whose speed $f_i(n_i)$ depends on the number of customers at node i, i.e. the instantaneous service completion rate at node i is $\mu_i f_i(n_i)$.

Type 2 node

In a type 2 node, there is a single server and the service discipline is processor sharing. That is, when there are *n* customers at a type 2 node, each is receiving service at the rate of $1/n$ of the service rate. The service times for class r customers are Coxian with parameters a_{irl}, b_{irl}, μ_{irl}, and L_{ir}. The node state S_i is defined as the vector $(v_1,v_2,...,v_R)$ where $v_r=(n_{ir1}, n_{ir2},...,n_{irL_{ir}})$ is a vector whose m-th element n_{irm} denotes the number of class r customers at node i which are in the m-th stage of their service. The speed of service at node i may depend on the total number of customers at node i as for type 1 nodes.

Type 3 node

In a type 3 node, there are as many servers as there may be customers requiring service (i.e., infinite server). As soon as a customer arrives, a separate server is assigned for the duration of the customer's service. The assumptions regarding the required service time distributions and the definition of the node state S_i are the same as for type 2 nodes.

Type 4 node

A single server is scheduled according to the preemptive resume last come first served (LCFS) discipline. In this discipline, a customer in service is preempted by an arriving customer. That is, when a new customer arrives the service of the current customer is interrupted until the new customer departs (which, in turn, may be interrupted) and then resumed from the point of interruption. The service time is Coxian, as for type 2 and 3 nodes. The state of the node, S_i, is defined as the vector of pairs $\{(r_1,m_1),(r_2,m_2),...,(r_{n_i}m_{n_i})\}$ where r_j and m_j are respectively the class index and the stage of service of the j-th customer in the LCFS order. The speed of the server may depend on the total number of customers at node i, as for type 1 nodes.

Let $M_r(S_i)$ be the number of class r customers at node i when the node is in state S_i. Then, network state S is feasible if

$$0 \leq M_r(S_i) \leq N_r, \ i=1,\ldots,M \quad \text{and} \quad \sum_{1 \leq i \leq M} M_r(S_i) = N_r, \ r=1,\ldots,R.$$

The steady state queue length distribution $\pi(S)$ is the solution of the global balance equations together with the normalizing equation:

$$\pi(S) \ [\text{instantaneous transition rate out of state } S] =$$
$$\sum \pi(S') \ [\text{instantaneous transition rate from } S' \text{ to } S]$$
$$\sum_S \pi(S) = 1 \tag{1.4}$$

The visit ratio, x_{ir}, is defined as the *relative* arrival rate of class r customers to node i. x_{ir}'s are determined from the following system of equations:

$$x_{ir} = \sum_{1 \leq j \leq M} x_{jr} p_{ji;r} \quad i=1,\ldots,M, \ r=1,\ldots,R. \tag{1.5}$$

For each class r, there are M unknowns (x_{ir}'s) and M equations of which M-1 are linearly independent. Hence, it is necessary to set one of the unknowns to an arbitrary value and calculate the others relative to the set value.

The following result is known as the BCMP theorem:

Theorem: Let e_{ir} be any solution of (1.5). The general solution of the global balance equations (1.4) has the form

$$\pi(S) = G^{-1} \prod_{1 \leq i \leq M} g_i(S_i) \tag{1.6}$$

where:

G: is the normalization constant which ensures that $\sum_S \pi(S) = 1$.

The factor $g_i(S_i)$ depends on the type of node i:

if node i is of type 1 then $g_i(S_i) = \displaystyle\prod_{j=1}^{n_i} \left[x_{ir_j} / (\mu_i f_i(j)) \right]$

if node i is of type 2 then

$$g_i(S_i) = n_i! \left\{ \prod_{r=1}^{R} \prod_{l=1}^{L_{ir}} \left[\left(x_{ir} A_{irl} / \mu_{irl} \right)^{n_{irl}} / n_{irl}! \right] \right\} \prod_{j=1}^{n_i} f_i(j)$$

if node i is of type 3 then $g_i(S_i) = \displaystyle\prod_{r=1}^{R} \prod_{l=1}^{L_{ir}} \left[\left(x_{ir} A_{irl} / \mu_{irl} \right)^{n_{irl}} / n_{irl}! \right]$

$$\text{if node } i \text{ is of type 4 then} \quad g_i(S_i) = \prod_{j=1}^{n_i} \left[x_{ir_j} A_{ir_j l_j} \Big/ \left(\mu_{ir_j l_j} f_i(j) \right) \right]$$

where $A_{irl} = \prod_{s=1}^{l-1} a_{irs} \, b_{irl}$.

As the steady state queue length distribution is the product of functions of nodes, these types of solutions to the global balance equations are referred to as product form solutions.

1.3.2 Numerical solutions

Any Markovian model can, in theory, be solved numerically. In particular, obtaining the steady state queue length distribution, π, of a queueing network is a three-step procedure:

i)   Determine the states and the state space of the network.
ii) Determine its state transition structure to construct the rate matrix, Q, of the network.
iii) Solve the linear system of equations $Q\pi = b$ numerically.

There are, however, some practical limitations in obtaining the steady state queue length distribution of queueing networks numerically. First, the state space of queueing networks grows rapidly with the number of nodes and the number of customers in the network. For example, the state space of a single-class closed network with ten type-1 nodes and ten customers in it has 1,847,560 states. Hence, the space required to store the rate matrix Q can be excessive, although this problem, to some extent, can be alleviated by utilizing the sparseness of Q (i.e. storing only the non-zero elements of Q). Second, the construction of Q from a model is rather a time-consuming task. Finally, solving the system of equations, in general, has a time complexity of $O(n^3)$ where n is the number of states, which restricts the applicability of numerical techniques.

1.3.3 Simulation

Simulations may be used to obtain the steady state queue length distributions of queueing networks. However, simulations could be considered an approximation technique. In particular, the exact values of the steady state queue length probabilities of a queueing network can be obtained if the network has a product form queue length distribution or a tractable numerical solution. However, the values obtained from simulation will, hopefully, be near (but usually not equal to) the exact values. Two of the principal problems with simulations are: determining how close simulation estimates are to the exact values, and determining how long to run the simulation in order to obtain estimates near the correct values. Furthermore, developing and implementing simulation models are usually a time consuming task.

1.3.4 Approximate solutions

Approximate solution techniques have been proposed to analyze queueing networks in which obtaining the exact solutions of the performance measures are inordinately expensive or the form of their steady state queue length distributions are not known. The main challenge with approximations is to bound the error in the solution. In general, the accuracy of an approximation is tested with numerical solutions (in smaller configurations that can be solved numerically) or with simulations to determine the conditions under which the algorithm yields a good approximation. Decomposition methods are the most widely used techniques in the approximate analysis of queueing networks. The main idea is to decompose the network into sub-networks, analyze each sub-network in isolation, and use the results obtained from each subsystem to analyze the macro-system composed of these subsystems. These methods give exact solutions for queueing networks with product form steady state queue length distributions. In networks with non-product form solutions, the method yields good approximations if the rate of interaction among the nodes in the sub-network is significantly higher than the rate of interaction of the sub-network with the remainder of the network. The error bounds for this approach can be calculated from the rate matrix of the network.

1.3.5 Performance metrics

To understand and to predict the behavior of a real life system, an analyst would like to know the values of the performance metrics of the system such as the percentage of time a device is utilized, the rate at which the system produces an output, the average unfinished work at each device, average time it takes to produce a unit at each device, average time it takes to produce a finished product, etc. Corresponding to these, the primary performance measures of queueing networks are defined as follows:

    i)   marginal queue length distribution,
    ii) utilization of each server,
    iii) throughput of each node,
    iv) throughput of the network,
    v) mean queue lengths,
    vi) average response time at node i.

The steady state marginal queue length probability, $\pi_i(n)$, of node i is the steady state probability of having n customers at node i, independent of the number of customers at other nodes of the network. Let S denote a network state, $\pi(S)$ be the steady state queue length probability of being in state S, and $Y_i(n)$ be the set of states such that there are n customers at node i. Then:

$$\pi_i(n) = \sum_{S \in Y_i(n)} \pi(S) \qquad n=0,1,...,N \qquad (1.7)$$

Let us consider a closed queueing network with N customers. The utilization, U_i, of node i is defined as the percentage of time the server is busy serving. In

Let us consider a closed queueing network with N customers. The utilization, U_i, of node i is defined as the percentage of time the server is busy serving. In networks with infinite node capacities, this is equivalent to the probability that the node is not empty, i.e. there is at least one customer at node i.

The throughput of node i is defined as the rate at which customers depart from that node. Let X_i be the throughput of node i. Then for a single server node i:

$$X_i = [1-\pi_i(0)]\, \mu_i \qquad\qquad (1.8)$$

where μ_i is the service rate at node i. In single class networks, the equations for the visit ratios, x_i, defined in (1.5) reduces to the following:

$$x_i = \sum_{1 \le j \le M} x_j p_{ji} \qquad i=1,...,M \qquad (1.9)$$

In the case of closed networks, there are M unknowns and M-1 independent equations in the above system. Hence, it is necessary to set one of the x_i's to one (or any other value) and solve the other x's relative to that given value. Without loss of generality, let the visit ratio of node 1 be set to one. In open networks, the set of equations is linearly independent. The throughput of a closed network, X, is defined as the average number of customers leaving node 1 per unit time. It is the number of customers leaving the network per unit time in the case of open network. The relation between the throughput of each node and the throughput of the network is given as follows:

$$X_i = x_i\, X \qquad\qquad i=1,...,M \qquad (1.10)$$

The mean queue length L_i of node i is the average number of customers in node i at steady state. It is given as:

$$L_i = \sum_{1 \le n \le N} n\, \pi_i(n) \qquad\qquad (1.11)$$

The average response time, T_i, is the average time a customer spends in node i at steady state. There is a simple relation between the average response time and the mean queue length in a queueing system in steady state that equates the average arrival rate to the average departure rate. In particular, since a customer remains at node i for an average time of T_i, his departure rate is $1/T_i$. The average number of customers at node i is L_i. Hence, the average departure rate is L_i/T_i. In steady state, the average arrival rate to node i, λ_i, is equal to the average departure rate from node i, hence, we have the following result which is known as the Little's result.

$$L_i = \lambda_i\, T_i \qquad\qquad (1.12)$$

We shall define with details performance metrics of queueing networks with blocking in the next chapter.

1.4 QUEUEING NETWORKS WITH FINITE CAPACITY QUEUES

Almost all queueing networks with product form queue length distributions require infinite queues, that is, it is assumed that there is always a space in the queue for arriving customers. In real life systems, the storage space is always finite. Hence, a more realistic modeling of systems with finite resources requires incorporation of finite node capacities into their queueing models. An important feature of queueing networks with finite queues is that the flow of customers through a node may be momentarily stopped when another node in the network reaches its capacity. That is, a phenomenon called *blocking* occurs. Queueing networks with blocking are difficult to solve: in general, their steady state queue length distributions could not be shown to have product form solutions. Hence, most of the techniques that are employed to analyze these networks are in the form of approximations, simulation, and numerical techniques.

In addition to the problem of blocking, deadlocks may occur in queueing networks with finite queues. In particular, a set of nodes is said to be in a deadlock state when every node in the set is waiting for a space to become available at another node in the set. In this case, all servers in the set are blocked and they can never get unblocked because the space required for the change of status of the server will never be available.

This book is a survey of analytical, approximate and numerical results related to queueing networks with finite queues and of their properties. Except for a few special cases, these networks could not be shown to have product form solutions. The steady state queue length distributions of these networks can, in theory, be calculated by solving the global balance equations together with the normalization equation numerically. In practice, however, this procedure can be restrictive due to the time complexity of the procedure and the large storage required to store the rate matrices, particularly for large networks. Since exact values of their steady state queue length distributions are, in general, not attainable, good approximation algorithms are required to analyze closed queueing networks with finite queues.

Chapter 2 introduces the concept of single class and multi-class queueing networks with finite capacities, blocking mechanisms, and the definition of performance indices. Various examples of real life systems modeled with queueing networks with blocking are illustrated in chapter 3. The exact analysis of queueing networks with blocking is presented in chapter 4 whereas special cases of queueing networks with exact product form solutions and of symmetrical networks are introduced in chapter 5. Chapter 6 is a summary of various approximation algorithms reported in the literature. Various properties including equivalences and monotonicity of queueing networks with blocking are discussed in chapter 7. The last chapter introduces the problem of buffer allocation in queueing networks.

1.5   BIBLIOGRAPHICAL NOTES AND REFERENCES

The Kendall's notation to describe the queueing process of a single queueing system was introduced in Kendall (1953). Gross and Harris (1974) present the fundamentals of queueuing theory. Cox (1955) introduced the representation of arbitrary distribution as a network of exponential stages as described in Section 2.1. The analysis of single queue is presented in Cohen (1969) and Kleinrock (1975) that also deals with queueing networks. Little's law given by formula (1.13) is presented in Little (1961).

Lavenberg (1983) and Kant (1992) present queueing network models, several methods for their analysis and their application for performance evaluation of computer and communication systems. An example of queueing network analysis to model a window flow control system is given in Reiser (1979).

After Jackson's pioneering work on product form queueing networks (1963), the development and analysis of one of the most general queueing networks was due to the combined efforts of Baskett, Chandy, Muntz and Palacios (1975). Their result is known as the BCMP theorem, presented in Section 1.3.1.

For numerical solution of Markovian models the reader interested in details on solving systems of linear equations may refer to Jennings (1977).

An overview of simulation of queueing networks is presented in Lavenberg (1983) and Chandy and Sauer (1978). The interested reader may also refer to Law and Kelton (1982) and Solomon (1983) for the details on modeling and analysis of queueing networks with simulation.

Approximate solution based on the decomposition principle was introduced by Courtois (1977) that evaluate error bounds on the state probability by analyzing the rate matrix of the network. Network decomposition is also discussed in Muntz (1978). The reader interested in approximate solution of queueing networks based on decomposition and aggregation methods may refer to Chandy and Sauer (1978) and Chandy, Herzog, and Woo (1975).

Next three sections list various books, survey and tutorials on queueing theory and queueing networks and finally references on queueing networks, mostly with blocking. Surveys on closed queueing networks with blocking can be found in Onvural (1990) and Balsamo and De Nitto Personè (1994). A survey on open queueing networks with blocking is presented in Perros (1989) and (1994).

1.5.1   Books on queueing theory and queueing systems

Cohen, J.W *The Single Server Queue*, North Holland Publishing Company, Amsterdam, 1969.

Kant, K. *Introduction to Computer System Performance Evaluation*. McGraw-Hill, 1992.

Kelly, F.P. *Reversibility and Stochastic Networks*. Wiley, 1979.

Kleinrock, L. *Queueing Systems. Vol.1 :Theory*. Wiley, 1975.

Lavenberg, S.S. *Computer Performance Modeling Handbook*. Prentice Hall, 1983.

1.5.2 Surveys, tutorials, and books on queueing networks with blocking

Akyildiz, I.F. and H.G. Perros Special Issue on Queueing Networks with Finite Capacity Queues, Performance Evaluation, Vol. 10, 3 (1989).

Balsamo, S. "Properties and analysis of queueing network models with finite capacities", in *Performance Evaluation of Computer and Communication Systems* (L.Donatiello, R.Nelson Eds.) Lecture Notes in Computer Science, 729, 1994, Springer-Verlag.

Balsamo, S. and V. De Nitto Personè "A survey of Product-form Queueing Networks with Blocking and their Equivalences" Annals of Operations Research, vol. 48 (1994) 31-61.

Onvural, R.O "Survey of Closed Queueing Networks with Blocking" ACM Computing Surveys, Vol. 22, 2 (1990) 83-121.

Onvural, R.O. Special Issue on Queueing Networks with Finite Capacity, Performance Evaluation, Vol. 17, 3 (1993).

Perros, H.G. "Open queueing networks with blocking" in *Stochastic Analysis of Computer and Communications Systems* (Takagi Ed.) North Holland, 1989.

Perros, H.G. *Queueing networks with blocking*. Oxford University Press, 1994.

1.5.3 Publications on queueing networks

Akyildiz, I.F. "Exact Product Form Solutions for Queueing Networks with Blocking" IEEE Trans. on Computers, Vol. 1 (1987) 121-126.

Akyildiz, I.F. "General Closed Queueing Networks with Blocking" in *Performance'87*, (P.J. Courtois and G. Latouche Eds.), 282-303, Elsevier Science Publishers, North Holland, Amsterdam, 1988.

Akyildiz, I.F. "On the Exact and Approximate Throughput Analysis of Closed Queueing Networks with Blocking" IEEE Trans. on Software Engineering, Vol. 14 (1988), 62-71.

Akyildiz, I.F. "Mean Value Analysis for Blocking Queueing Networks" IEEE Trans. on Software Engineering, Vol. 14 (1988) 418-429.

Akyildiz, I.F. "Product Form Approximations for Queueing Networks with Multiple Servers and Blocking" IEEE Trans. Computers, Vol. 38 (1989) 99-115.

Akyildiz, I.F. "Analysis of Queueing Networks with Rejection Blocking" in Proc. First International Workshop on Queueing Networks with Blocking, (H.G. Perros and T. Altiok Eds.), Elsevier Science Publishers, North Holland, Amsterdam, 1989.

Akyildiz, I.F., and S. Kundu "Deadlock Free Buffer Allocation in Closed Queueing Networks" Queueing Systems Journal, 4 (1989) 47-56.

Akyildiz, I.F., and J. Liebeherr "Application of Norton's Theorem on Queueing Networks with Finite Capacities" in Proc. INFOCOM 89, 914-923, 1989.

Akyildiz, I.F., and H. Von Brand "Duality in Open and Closed Markovian Queueing Networks with Rejection Blocking" Tech. Rep., CS-87-011, (1987), Louisiana State University.

Akyildiz, I.F., and H. Von Brand "Exact solutions for open, closed and mixed queueing networks with rejection blocking" J. Theor. Computer Science, 64 (1989) 203-219.

Akyildiz, I.F., and H. Von Brand "Computation of Performance Measures for Open, Closed and Mixed Networks with Rejection Blocking" Acta Informatica (1989)

Akyildiz, I.F., and H. Von Brand "Central Server Models with Multiple Job Classes, State Dependent Routing, and Rejection Blocking" IEEE Trans. on Softw. Eng., Vol. 15 (1989) 1305-1312.

Allen, A.O. *Probability, Statistics and Queueing Theory with Computer Science Applications.* Academic Press, New York, 1990.

Altiok, T., and H.G. Perros "Approximate Analysis of Arbitrary Configurations of Queueing Networks with Blocking" Annals of Oper. Res. 9 (1987) 481-509.

Ammar, M.H., and S.B. Gershwin "Equivalence Relations in Queueing Models of Fork/Join Networks with Blocking" Performance Evaluation, Vol. 10 (1989) 233-245.

Balsamo, S., and C. Clò "A Convolution Algorithm for Product Form Queueing Networks with Blocking" Annals of Operations Research, Vol. 79 (1998) 97-117.

Balsamo, S., V. De Nitto Personè, and G. Iazeolla "Identity and Reducibility Properties of Some Blocking and Non-Blocking Mechanisms in Congested Networks" in *Flow Control of Congested Networks*, (A.R. Odoni, L. Bianco, G. Szego Eds.), NATO ASI Series, Comp. and System Science, Vol.F38, Springer-Verlag, 1987.

Balsamo, S., and V. De Nitto Personè "Closed queueing networks with finite capacities: blocking types, product-form solution and performance indices" Performance Evaluation, Vol. 12, 4 (1991) 85-102.

Balsamo, S., and V. De Nitto Personè "A survey of Product-form Queueing Networks with Blocking and their Equivalences" Annals of Operations Research, vol. 48 (1994) 31-61.

Balsamo, S., and L. Donatiello "Two-stage Queueing Networks with Blocking: Cycle Time Distribution and Equivalence Properties", in *Modelling Techniques and Tools for Computer Performance Evaluation* (R. Puigjaner, D. Potier Eds.) Plenum Press, 1989.

Balsamo, S., and L. Donatiello "On the Cycle Time Distribution in a Two-stage Queueing Network with Blocking" IEEE Transactions on Software Engineering, Vol. 13 (1989) 1206-1216.

Balsamo, S., and G. Iazeolla "Some Equivalence Properties for Queueing Networks with and without Blocking" in *Performance '83* (A.K. Agrawala, S.K. Tripathi Eds.) North Holland, 1983

Baskett, F., K.M. Chandy, R.R. Muntz, and G. Palacios "Open, closed, and mixed networks of queues with different classes of customers" J. of ACM, 22 (1975) 248-260.

Boxma, O., and A.G. Konheim "Approximate analysis of exponential queueing systems with blocking" Acta Informatica, 15 (1981) 19-66.

Caseau, P., and G. Pujolle "Throughput capacity of a sequence of transfer lines with blocking due to finite waiting room" IEEE Trans. on Softw. Eng. 5 (1979) 631-642.

Chandy, K.M., U. Herzog, and L. Woo "Parametric analysis of queueing networks" IBM J. Res. Dev., 1 (1975) 36-42.

Chandy, K.M., and C.H. Sauer "Approximate Methods for Analyzing Queueing Network Models of Computing Systems" ACM Computing Surveys Vol. 10 (1978) 281-317.

Coffman, E.G., M.J. Elphick, and A. Shoshani "System Deadlocks" ACM Computing Surveys, Vol. 2 (1971) 67-78.

Cohen, J.W *The Single Server Queue*, North Holland Publishing Company, Amsterdam, 1969.

Courtois, P. J. *Decomposability: Queueing and Computer System Applications*, Academic Press, Inc, New York, 1977.

Cox, D.R. "A Use of Complex Probabilities in the Theory of Stochastic Processes" in Proc. Cambridge Philosophical Society, 51 (1955) 313-319.

Dallery, Y., and Y. Frein "A decomposition method for the approximate analysis of closed queueing networks with blocking", Proc. First Int. Workshop on Queueing Networks with Blocking, (H.G. Perros and T. Altiok Eds.) North Holland, 1989.

Dallery, Y., and D.D. Yao "Modelling a system of flexible manufacturing cells" in: Modeling and Design of Flexible Manufacturing Systems (Kusiak Ed.) North-Holland (1986) 289-300.

De Nitto Personè, V. "Topology related index for performance comparison of blocking symmetrical networks" European J. of Oper. Res., Vol. 78 (1994) 413-425.

De Nitto Personè, V., and D. Grillo "Managing Blocking in Finite Capacity Symmetrical Ring Networks" in Proc. 3rd Conference on Data and Communication Systems and Their Performance, Rio de Jenerio, Brasil, 1987.

De Nitto Personè, V., and V. Grassi "An analytical model for a parallel fault-tolerant computing system" Performance Evaluation, Vol. 38 (1999) 201-218.

Diehl, G.W. "A Buffer Equivalency Decomposition Approach to Finite Buffer Queueing Networks" Ph.D. thesis, Eng. Sci., Harvard University, 1984.

Gelenbe, E., and I. Mitrani. *Analysis and Synthesis of Computer Systems*. Academic Press, Inc., London, 1980.

Gershwin, S., and U. Berman "Analysis of Transfer Lines Consisting of Two Unreliable Machines with Random Processing Times and Finite Storage Buffers" AIIE Trans., 1Vol. 3- (1981) 2-11.

Gordon, W.J., and G.F. Newell "Cyclic queueing systems with restricted queues" Oper. Res., Vol. 15 (1967) 286-302.

Gordon, W.J., and G.F. Newell "Cyclic queueing systems with Exponential Servers" Oper. Res., Vol. 15 (1967) 254-265.

Gross, D., and C.M. Harris. *Fundamentals of Queueing Theory*. John Wiley and Sons, Inc., New York, 1974.

Highleyman, W. H. *Performance Analysis of Transaction Processing Systems*. Prentice Hall, Inc., Englewood Cliffs, New Jersey, 1989.

Hordijk, A., and N. Van Dijk "Networks of queues with blocking", in: Performance '81 (K.J. Kylstra Ed.) North Holland, 1981, 51-65.

Hordijk, A., and N. Van Dijk "Adjoint Processes, Job Local Balance and Insensitivity of Stochastic Networks", Bull: 44 session, Int. Stat. Inst., Vol. 50 (1982) 776-788.

Hordijk, A. , and N. Van Dijk "Networks of Queues: Part I-Job Local Balance and the Adjoint Process" ; "Part II-General Routing and Service Characteristics" in Proc. Int. Conf. Modeling Comput. Sys., 60 (1983) 158-205.

Jackson, J.R. "Jobshop-like Queueing Systems" Mgmt. Sci., Vol. 10 (1963) 131-142.

Jennings, A. *Matrix Computation for Engineers and Scientists*. John Wiley and Sons, Inc., New York, 1977.

Jun, K.P. "Approximate Analysis of Open Queueing Networks with Blocking", Ph.D. thesis, Operations Research Program, North Carolina State University, 1988.

Kelly, K.P. *Reversibility and Stochastic Networks*. John Wiley and Sons Ltd., Chichester, England, 1979.

Kendall, D. G. "Stochastic Processes Occurring in the Theory of Queues and Their Analysis by the Method of Imbedded Markov Chains" Ann. Math. Statist., 24 (1953) 338-354

Kleinrock, L. *Queueing Systems-Vol. II*, John Wiley and Sons, New York, 1976.

Kouvatsos, D.D. "Maximum Entropy Methods for General Queueing Networks" in Proc. Modeling Tech. and Tools for Perf. Analysis, (Potier Ed.), North Holland, Amsterdam, 1983, 589-608.

Kouvatsos, D.D., and N.P., Xenios "Maximum Entropy Analysis of General Queueing Networks with Blocking", in First International Workshop on Queueing Networks with Blocking, Perros and Altiok (Eds), Elsevier Science Publishers, North Holland, 1989.

Law, A.M., and W.D. Kelton *Simulation Modeling and Analysis*. Mc Graw Hill, New York, 1982.

Little, J.D.C. "A Proof of the Queueing Formula L=λW" Oper. Res., Vol. 9 (1961) 383-387

Marie, R. "An Approximate Analytical Method for General Queueing Networks" IEEE Trans. on Software Engineering, Vol. 5 (1979) 530-538.

Minoura, T. "Deadlock Avoidance Revisited" Journal of ACM, Vol. 29 (1982) 1023-1048.

Mitra, D., and I. Mitrani "Analysis of a Novel Discipline for Cell Coordination in Production Lines" AT&T Bell Labs Res. Rep., 1988.

Muntz, R.R. "Queueing Networks: A Critique of the State of the Art Directions for the Future" ACM Comp. Surveys, Vol. 10 (1978) 353-359.

Onvural, R.O. "Closed Queueing Networks with Finite Buffers" Ph.D. thesis, CSE/OR, North Carolina State University, 1987.

Onvural, R.O. "On the Exact Decomposition of Exponential Closed Queueing Networks with Blocking" in First International Workshop on Queueing Networks with Blocking, (H.G. Perros and T. Altiok Eds), Elsevier Science Publishers, North Holland, 1989.

Onvural, R.O. "Some Product Form Solutions of Multi-Class Queueing Networks with Blocking' Performance Evaluation, Special Issue on Queueing Networks with Blocking, Akyildiz and Perros (Eds), 1989.

Onvural, R.O., and H.G. Perros "On Equivalencies of Blocking Mechanisms in Queueing Networks with Blocking" Oper. Res. Letters, Vol. 5 (1986) 293-298.

Onvural, R.O., and H.G. Perros "Equivalencies Between Open and Closed Queueing Networks with Finite Buffers" in Proc. International Seminar on the Performance Evaluation of Distributed and Parallel Systems, Kyoto, Japan. Also to appear in Performance Evaluation, 1988.

Onvural, R.O., and H.G. Perros "Some equivalencies on closed exponential queueing networks with blocking" Performance Evaluation, Vol.9 (1989) 111-118.

Onvural, R.O., and H.G. Perros "Throughput Analysis in Cyclic Queueing Networks with Blocking" IEEE Trans. Software Engineering, Vol. 15 (1989) 800-808.

Perros, H.G. "Queueing Networks with Blocking: A Bibliography" ACM Sigmetrics, Performance Evaluation Review, Vol. 12 (1984) 8-12.

Perros, H.G. "Open Queueing Networks with Blocking" in *Stochastic Analysis of Computer and Communications Systems*, (Takagi Ed.), Elsevier Science Publishers, North Holland, 1989.

Perros, H.G., A. Nilsson, and Y.C. Liu "Approximate Analysis of Product Form Type Queueing Networks with Blocking and Deadlock" Performance Evaluation (1989).

Pittel, B. "Closed Exponential Networks of Queues with Saturation: The Jackson Type Stationary Distribution and Its Asymptotic Analysis" Math. Oper. Res., Vol. 4 (1979) 367-378.

Reiser, M. "A Queueing Network Analysis of Computer Communications Networks with Window Flow Control" IEEE Trans. on Comm., Vol. 27 (1979) 1199-1209.

Sereno, M. "Mean Value Analysis of product form solution queueing networks with repetitive service blocking" Performance Evaluation, Vol. 36-37 (1999) 19-33.

Shanthikumar, G.J., and D.D. Yao "Monotonicity Properties in Cyclic Queueing Networks with Finite Buffers" in First International Workshop on Queueing Networks with Blocking, (H.G. Perros and T. Altiok Eds), Elsevier Science Publishers, North Holland, 1989.

Solomon, S.L. *Simulation of Waiting Line Systems*. Prentice Hall, Englewood Cliffs, New Jersey, 1983.

Suri, R., and G.W. Diehl "A New Building Block for Performance Evaluation of Queueing Networks with Finite Buffers" in Proc. ACM Sigmetrics on Measurement and Modeling of Computer Systems (1984) 134-142.

Suri, R., and G.W. Diehl "A Variable Buffer Size Model and Its Use in Analytical Closed Queueing Networks with Blocking" Management Science, Vol. 32 (1986) 206-225.

Trivedi, K.S. *Probability and Statistics with Reliability, Queueing and Computer Science Applications*. Prentice Hall, Englewood Cliffs, New Jersey, 1982.

Van Dijk, N.M., and H.C. Tijms "Insensitivity in Two Node Blocking Models with Applications" in *Teletraffic Analysis and Computer Performance Evaluation* (Boxma, Cohen and Tijms Eds.), Elsevier Science Publishers, North Holland, 1986, 329-340.

Whitt, W. "Open and Closed Models for Networks of Queues" AT&T Bell Labs Tec. J., Vol. 63 (1984) 1911-1979.

Yao, D.D., and J.A. Buzacott "Modeling a Class of State Dependent Routing in Flexible Manufacturing Systems" Annals of Operations Research, Vol. 3 (1985) 153-167.

Yao, D.D., and J.A. Buzacott "Queueing Models for Flexible Machining Station Part I: Diffusion Approximation" Eur. J. Operations Research, Vol. 19 (1985) 233-240.

Yao, D.D., and J.A. Buzacott "Queueing Models for Flexible Machining Stations Part II: The Method of Coxian Phases" Eur. J. Operations Research, Vol. 19 (1985) 241-252.

Yao, D.D., and J.A. Buzacott "The Exponentialization Approach to Flexible Manufacturing System Models with General Processing Times" Eur. J. of Operations Research, Vol. 24 (1986) 410-416.

2 QUEUEING NETWORKS WITH BLOCKING

In this chapter, we introduce queueing networks with finite capacity queues. When limitations are imposed on the queue capacities, a phenomenon called blocking occurs. Simply defined, blocking is forcing a departure from a queue or an arrival to a queue to stop temporarily due to lack of space in the queue. In Section 2.1 we define single class networks. There are different blocking mechanisms introduced in the literature to model different types of flow systems with finite resources. These blocking mechanisms are defined in Section 2.2. The relation between blocking and state dependent routing is introduced in Section 2.3. The extension of results presented in the context of single class networks with finite capacities to multiclass networks with blocking is presented in Section 2.4. Finally, various performance metrics of interest in the analysis of blocking networks are defined in Section 2.5. Bibliographical notes and references are presented in Section 2.6.

2.1 SINGLE CLASS NETWORKS

Consider a queueing network consisting of M service centers (or nodes). The number of customers waiting in one or more service queues is restricted due to finite holding space associated with these nodes. The system may be modeled as an open or closed queueing network with finite capacities. When modeled as a closed network, the total number of customers waiting for a service and those currently receiving service is defined by the parameter N. Open queueing networks require the definition of exogenous arrival process(es). In this case the arrival rate can depend on the number of customers in a node/network (i.e., load dependent arrival process) or may be independent on the population in the network. The notation used to denote load independent and dependent arrival rates are defined, λ and λ a(n), n$\geq$0, respectively, where a(n) is an arbitrary non-negative function that depends on the total number of customers in a node/network.

An exogenous arrival chooses to enter node i with probability p_{0i}, 1$\leq$i$\leq$M. Hence, the arrival rate at node i is defined as λ p_{0i} for load independent arrivals, and λ a(n) p_{0i} for load dependent arrivals.

A customer leaving node i attempts to enter node j with probability p_{ij}, 1$\leq$i,j$\leq$M. P= [p_{ij}] (1$\leq$i,j$\leq$M) denotes the matrix of routing probabilities. More generally, routing matrix P can be defined as a function of the state of the network. Without loss of generality, the formulation presented next is based on state independent routing matrix. We discuss the use of state dependent routing to model finite capacity queues and blocking in Section 2.3.

For open networks p_{i0} ($1 \leq i \leq M$) denotes the probability that a job leaves the network upon completing its service at node i. The following relation holds, by definition, $1 \leq i \leq M$:

$$\sum_{j=1}^{M} p_{ij} + p_{i0} = 1$$

Let us introduce vector $\mathbf{x} = (x_1, \ldots, x_M)$ which can be obtained by solving the following linear system:

$$x_i = \lambda p_{0i} + \sum_{j=1}^{M} x_j p_{ji} \qquad (2.1)$$

Equation (2.1) is referred to as traffic balancing equations. In closed networks, we have, $p_{0i} = p_{i0} = 0$ for $1 \leq i \leq M$ since there are no arrival to and departures from the system. In this case, equation (2.1) does not provide a unique solution.

For open queueing networks with infinite capacity queues, x_i in equation (2.1) represents the throughput of node i, whereas it is referred to as the relative throughput or relative visit ratio of customers at node i for closed networks, $1 \leq i \leq M$. This is because equation (2.1) is solved in closed queueing networks by setting one of the x_i's to an arbitrary, non-negative value. The definition of x_i's does not hold in general for queueing networks with finite capacity queues.

Service center i is described by the number of servers, the service time distribution, the service discipline, and the queue capacity.

Let n_i denote the number of jobs at node i. K_i denotes the number of identical servers at node i. The service time distribution of jobs is denoted by $F_i(t)$, $t \geq 0$ with mean service time being defined by $1/\mu_i$ (load independent case). If the service rate at node i depends on the number of customers in its queue, $n_i \geq 0$, then the service rate of node i with K_i servers and n_i customers is defined by $\mu_i f_i(n_i, K_i)$, where f_i is an arbitrary non-negative function with $f_i(0, K_i) = 0$. For a single server node we denote this function as $f_i(n_i)$.

The service disciplines are defined based on a special framework. Let w_i denote the number of customers in the queue of node i and p denote the position of a job in node i, $1 \leq p \leq w_i$, $1 \leq i \leq M$. Furthermore, let $\delta(p, w_i)$ denote the probability that an arriving job is placed in position p when there are w_i jobs in the queue, $1 \leq p \leq w_i + 1$, and $\varphi_i(p, w_i)$ denote the fraction of service given to the job in position p, $1 \leq p \leq w_i$, $w_i \geq 0$, $1 \leq i \leq M$. By definition, we have:

$$\sum_{p=1}^{w_i+1} \delta(p, w_i) = \sum_{p=1}^{w_i} \varphi(p, w_i) = 1 \qquad (2.2)$$

A service discipline is said symmetric or station balancing if it satisfies the following condition:

$$\delta(p, w_i) = \varphi_i(p, w_i+1) \qquad 1 \le p \le w_i+1, \ w_i \ge 0 \qquad (2.3)$$

Service disciplines that can be defined based on this framework includes First Come First Service (FCFS), Last Come First Service pre-emptive resume (LCFS), Processor Sharing (PS) and Random service. FCFS and Random disciplines are not symmetric, while LCFS and PS are symmetric policies. Any scheduling policy that depends on the service time or on job priority cannot be represented based on this framework.

Queueing networks with finite capacity queues include additional constraints on the maximum number of customers in the system where a system can be a single resource or it can be a (sub)network. Imposing these constraints to queueing models require additional parameters to be defined.

Let B_i denote the capacity of node i, i.e., the maximum number of customers admitted at node i, i.e., the total number of jobs at node i, n_i, is less than or equal to the capacity of the node. When the number of customers at a node reaches its capacity ($n_i = B_i$), the node is said to be full.

Similarly, let B_W denote the maximum population allowed in a subnetwork W. In this context, a sub-network refers to a subset of nodes of a network. It is possible to impose a minimum population constraint L_W for subnetwork W to model different types of flow systems. That is, $n_W = \Sigma_{i \in W} n_i$, such that $L_W \le n_W \le B_W$.

When limitations are imposed on the maximum number of customers that might be allowed in a node/subnetwork/network, the admission of new customers to a full node is stopped until a space becomes available. Customers who completed their service at a node but can not depart due to lack of space in their destination node are forced to wait in the node they completed their service (i.e., that is the only space available to hold the customer). Note that B_i denotes the total capacity of node i, that is the queue capacity and the server space if it can be used to hold blocked customer(s). Let $b_i(n_i)$ denote the probability that a job arriving at node i, is accepted when there are n_i customers in the system. $b_i(n_i)$ is referred to as the blocking function.

As an example of a simple blocking function, consider the case in which an arriving customer is blocked at a time its destination node is full. In this case, we have:

$$b_i(n_i) = 1 \qquad \text{for } 0 \le n_i < B_i, \qquad b_i(B_i) = 0 \qquad 1 \le i \le M$$

More generally, one can define

$$0 < b_i(n_i) \leq 1 \quad \text{for } 0 \leq n_i < B_i, \quad b_i(B_i) = 0 \quad 1 \leq i \leq M. \quad (2.4)$$

2.2   BLOCKING MECHANISMS

When limitations are imposed on the total number of customers waiting/receiving service in a queue, a new arrival can not enter the queue if it arrives at a time the queue is full. The arriving customer, in this case, is called "blocked". A blocking mechanism is a description of a blocked customer's behavior. We consider five different blocking mechanisms (or types) that describe different types of behaviours upon blocking defined in the literature.

The first three blocking types are named Blocking After Service, Blocking Before Service and Repetitive Service Blocking. The other two blocking mechanisms are named Stop and Recirculate Blocking.

Each blocking mechanism is formally defined next.

Blocking After Service (BAS): When a job upon completion of its service at node i attempts to enter to a full queue (i.e., node i is blocked), it is forced to wait at node i occupying the server space, until a space becomes available at destination node j. Server i is forced to stop processing jobs that might be waiting in its queue (it is blocked) until the blocked customer can enter its destination queue. Upon departure from node i the server space becomes available for another job waiting in the queue. In this case, service at node i is resumed as soon as a departure occurs from node j (at which time, the job waiting at node i moves to its destination node immediately).

Figure 2.1 illustrates BAS blocking at a time the customer at node i completed its service while node j is full and as a consequence node i is blocked.

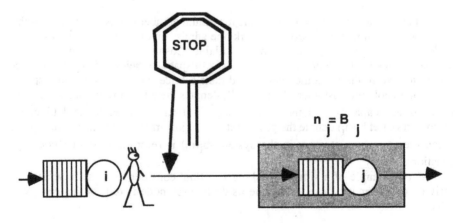

Figure 2.1. Blocking After Service

It is possible that the destination queue of customers receiving service in two or more different queues is the same. If these customers complete their service and they all find the destination queue full, more than one queue is blocked by the same queue. In this case, it is necessary to define the order in which blocked customers will be unblocked, as space becomes available in the blocking node. First Blocked First Unblocked discipline states that the first customer that joins the blocking queue when a departure occurs is the one that was blocked first.

BAS blocking mechanism has been used mainly to model production systems and disk I/O subsystems. It has also referred to as classical, transfer, manufacturing and production blocking in the literature.

Blocking Before Service (BBS): In this blocking mechanism, a job declares its destination node j before it starts receiving service at node i. The destination of the job does not change until it completes its service and departs from the queue. If upon entering service at node i, a job finds node j full, the service at node i does not start and server i is blocked. It is also possible that at time a job enters service, there was a space available at node j but becomes full prior to service completion at node i. When the destination node becomes full, the service at node i is interrupted and the server is blocked. The service at node i is resumed as soon as a departure occurs from node j. It is generally assumed that the amount of service a job has received until it is blocked is lost, i.e. a new sevice starts when a node becomes unblocked.

Figure 2.2 illustrates BBS blocking. The customer at server i is blocked if its destination (node j) is full upon entering the server or becomes full while it is receiving its service. Figure 2.3 illustrates a case in which service at node i is interrupted when node j becomes full due to a customer moving from node k to node j, while a job at node i is receiving service.

BBS blocking is further classified into two subcategories depending on whether or not the server space can be used to hold a job when the server is blocked:

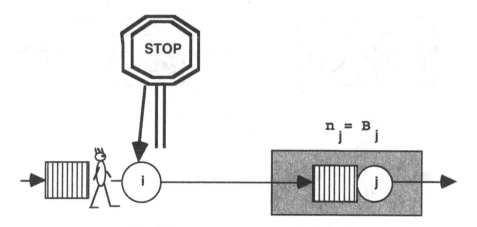

Figure 2.2. Blocking Before Service

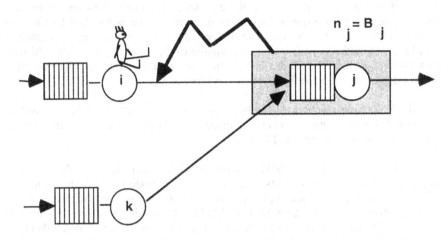

Figure 2.3. Service interruption in BBS

BBS-SO (server occupied): In this case, the service space is used to hold a customer during the time the server is blocked;

BBS-SNO (server is not occupied): In this case, the service space cannot be used to hold a customer during the time the server is blocked. BBS-SNO is not well defined for queueing networks with arbitrary topologies.

Consider the network in figure 2.4 at a time node i is full and there are B_j-1 customers at node j. When a service completion at node k occurs, node j becomes full and node i is blocked. Since node i is full at this time, there is no space to place

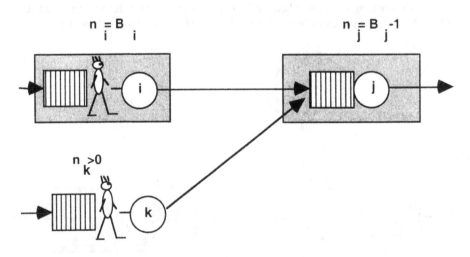

Figure 2.4. Non-feasible topology for BBS-SNO

the customer who is currently receiving service. Since BBS-SNO does not allow the server space to be used during blocking we observe an infeasible problem that arises due to the definition of the BBS-SNO blocking mechanism.

BBS-SNO is applicable only in network topologies in which a node with a finite capacity has only one upstream node with a finite capacity feeding it.

We can define a variant of the BBS blocking as follows.

BBS-O (Overall Blocking Before Service): If a node becomes full, it causes all its upstream nodes to be blocked, regardless of the destinations of jobs currently receiving service at each upstream node. The destination of a job is declared upon entering the server and does not change until the job departs from the queue. Similar to BBS-SO blocking, the server space is used to hold the blocked customer and services in each upstream node are resumed as soon as a departure occurs from node j.

BBS blocking mechanism has been used to model production, telecommunication, and computer systems. It has also referred to as service or immediate blocking in the literature.

Repetitive Service Blocking (RS): A job completes its service at node i and attempts to enter node j. If node j at that time is full, the job at node i starts receiving a new and independent service according to its service discipline.

Figure 2.5 illustrates RS blocking in which a customer is rejected by node j due to lack of buffer space and it starts receiving a new service at node i.

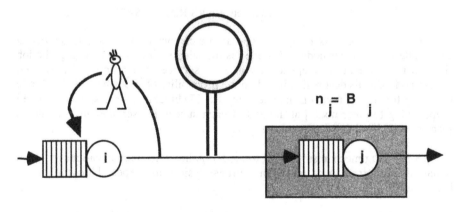

Figure 2.5. Repetitive Service Blocking

RS blocking is categorized into two subcategories depending on whether a blocked job chooses a new destination node independently of its previous selection(s) or it keeps trying to enter the same destination node at each service completion:

RS-RD (random destination): In this case, a job chooses its destination randomly every time it completes its service and finds the destination node full, independent of its previous choices;

RS-FD (fixed destination): In this case, a job declares its destination at the time it enters the server and does not change it until it departs from the queue, independent of the number of times it attempts to enter its destination node.

Figure 2.6 illustrates node i with RS blocking with two destination nodes j and k.

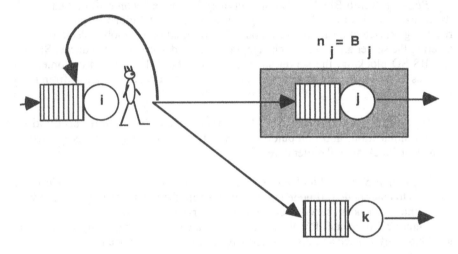

Figure 2.6. Difference between RS-RD and RS-FD

Consider the network in figure 2.6 at a time customer completing its service at node i attempts to enter node j. If node i is modeled with RS-RD blocking, the job chooses its destination independently at each service completion and may well choose node k as its destination based on the probability of choosing node k or node j as its destination. If node i is modeled with RS-FD blocking, the customer at node i keeps trying to enter node j at the end of each successive service, until it finds a space at node j and departs from node i.

RS blocking mechanism is used to model mainly telecommunications systems. It has also referred to as rejection, retransmission and repeat blocking in the literature.

In the following in some cases we simply denote by BBS and RS the two blocking types when we intend to include all the subtypes.

Remark. A different kind of blocking mechanism called *generalized blocking* can be defined where the server continue processing customers in the queue even if the destination node is full. The customers that have completed service at node i but cannot be sent to the next node, continue to share the buffer space of node i along with the other customers that are either waiting for service or being served upon. The customers arriving at this node when the queue is full are lost.

This blocking is also called *kanban blocking* due to its application to modeling kanban-controlled manufacturing systems. Under this blocking mechanism a server continue the service as long as there are customers available, even if the destination node is full. Customers can be raw or completed. A raw customer has not yet been served and a completed customer has already been served by the node. The buffer at node i is divided in two portions, one for raw customers with capacity α_i and the other for the finished customers with capacity β_i. If the buffer for raw customers at the destination node is full, then customers wait in buffer β_i at node i. The total queue capacity of node i is B_i. Hence the following relations hold: $0\leq\alpha_i$, $\beta_i\leq B_i$ and $\alpha_i+\beta_i\leq B_i$. Therefore a customer completing service at node i can be blocked and forced to wait in buffer β_i if the destination buffer α_j or the total capacity of node j is the maximum allowed B_j. The server of node i can be blocked if buffer β_i is full and it is starved if there are no raw customers in α_i. For particular values of the parameters α_i and β_i this blocking reduces to other blocking types. Specifically general blocking includes BBS-SO for $\alpha_i=B_i$ and $\beta_i=0$, BAS for $\alpha_i=B_i$ and $\beta_i=1$ for single server node or more genrally, $\beta_i=K_i$.

So far, we have defined the three types of blocking mechanisms BAS, BBS and RS, in which a server/job is blocked due to its destination being full. In the next two types of blocking mechanisms, the population in the network is assumed to be in the range [L,U], where L and U are the minimum and maximum populations admitted, respectively.

Let us first define a(n) to denote the load dependent arrival rate function and d(n) to denote a non-negative departure blocking function, where $n\geq0$ is the network population. We have:

$$a(n) = 0 \qquad \text{for } n\geq U$$
$$d(n) = 0 \qquad \text{for } n\leq L.$$

STOP Blocking: In this blocking mechanism, the service rate at each node depends on the number of customers in the network (n), according to the function d(n). When d(n) = 0, the service at each node is stopped. Service at a node is resumed upon arrival of a new customer to the network.

STOP blocking mechanism is used to model communications systems. It is also referred to as delay blocking in the literature.

RECIRCULATE Blocking: In this blocking mechanism, a job upon completion of its service at node i leaves the network with probability $p_{i0}\, d(n)$, when n is the total network population and it is forced to stay in the network with probability p_{i0} [1-d(n)], where p_{i0} is the routing probability. Consequently, a job completing the service at node i enters node j with state dependent routing probability $p_{ij} + p_{i0}$ [1-d(n)] p_{0j}, $1\leq i,j\leq M$, $n\geq0$.

This Recirculate blocking mechanism is used to model telecommunications systems and it is also referred to as triggering protocol in the literature.

Unlike networks with infinite queues, queueing networks with finite capacities are subject to deadlock depending on the blocking mechanism used to model the network. Consider two nodes and the topology of the network is such that there is a non-zero probability that a job will receive service at both nodes and come back to the node it first received a service. In this scenario, it is possible that a customer at node 1 is waiting for a space to become available at node 2 whereas a customer at node 2 is waiting for a space to become available at node 1. Since each is waiting for the other, a deadlock occurs if a space in one of the two nodes never becomes available.

If the network topology is such that a deadlock may occur, two possible approaches to addressing the problem are (i) preventive techniques, and (ii) detection and resolve techniques.

Deadlock prevention for blocking types BAS, BBS and RS-FD can be achieved by imposing the condition that the network population N is less than the total buffer capacities of nodes along each possible cycle in the network. In the case of RS-RD blocking, it is sufficient that routing matrix P is irreducible and N is less than the total buffer capacity of the nodes in the network.

More precisely, a network with RS-RD blocking is deadlock-free if

$$N < \sum_{j=1}^{M} B_j \qquad (2.5)$$

Let $c=(i_1,i_2,...,i_z)$ denote a cycle in the network with $i_1=i_z=i$ and routing probabilities $p_{i_j i_{j+1}} \neq 0$, $1 \leq j \leq z$. Then a network with BAS, BBS-SO, BBS-O and RS-FD blocking is deadlock-free if

$$N < \min_{\forall c} \sum_{i_j \in c} B_{i_j} \qquad (2.6)$$

Similarly, a network with BBS-SNO blocking is deadlock-free if

$$N < \min_{\forall c} \sum_{i_j \in c} (B_{i_j} - 1) \qquad (2.7)$$

The minimum in equations (2.6) and (2.7) is computed over all the possible cycles in the network.

In order to avoid deadlocks for BAS, BBS and RS-FD blocking types, it is necessary to assume that $p_{ii}=0$, $1 \leq i \leq M$. However, one can assume that a customer at node i cannot be blocked by i itself.

Finally, Stop and Recirculate blocking mechanisms themselves do not cause any deadlock.

2.3   BLOCKING AND STATE DEPENDENT ROUTING

In this section we consider queueing networks with finite capacity and state dependent routing.

Routing probabilities can be defined as a function of the network state. For example, given the state of a network with M nodes $\mathbf{n} = (n_1, n_2, ..., n_M)$, state dependent routing can be defined as an arbitrary function $P(\mathbf{n})=[p_{ij}(\mathbf{n})]$ $(1 \leq i,j \leq M)$. By defining the state dependent function appropriately, the probability that a job leaving node i enters node j can depend on the state of any one node, depend on the state of a subnetwork or on the state of the entire network.

Blocking mechanisms defined previously describe the behaviors of both the customers in the network and the servers. Next we will discuss how state dependent routing can be used to represent the behaviors of customers in a network with finite capacity queues. In this context, state dependent routing allows us to model more general job behaviors than those represented by the blocking mechanisms defined in Section 2.2. However, state dependent routing can not be used to describe the behaviors of servers in the network. In particular, not all these blocking mechanisms can be described with a state dependent routing function. As a special case, Recirculate blocking can be defined using state dependent routing as follows:

$$p_{ij}(\mathbf{n}) = p_{ij} + p_{i0} \, [1-d(n)] \, p_{0j}$$

Similarly, RS-RD blocking can be defined using the following state dependent routing:

$$p_{ij}(\mathbf{n}) = p_{ij} \, b_j(n_j)$$

where blocking function $b_j(n_j)$ was defined in Section 2.1.

On the contrary, state dependent routing cannot be used to model Stop, BAS and BBS blocking mechanisms.

In general, the description of a queueing network with finite capacity queues can include state dependent routing under different blocking mechanisms, thereby providing a more powerful tool than only one of the two can provide alone.

2.4   MULTICLASS QUEUEING NETWORKS WITH FINITE CAPACITY QUEUES

So far we have introduced single class queueing networks with finite capacities. In this section, we extend this model to include multiple classes of customers.

Consider a queueing network with finite capacities with C classes of customers. The set of job classes are partitioned into R disjoint sets $E_1,...,E_R$, referred to as chains. Each chain is either open or closed. In an open chain, customers belonging to the chain arrive from outside and depart from the network after receiving service at one or more nodes. In a closed chain, there is a fixed number of customers circulating among network nodes at all times. In an open network, all chains are open chains whereas in a closed network all chains are closed chains, respectively. It is also possible to construct a queueing network consisting of both open and closed chains. These networks are referred to as mixed networks.

In a closed chain r, there are N_r customers circulating in the network at all times, $1 \leq r \leq R$, and $N = \Sigma_{1 \leq r \leq R} N_r$ is the total network population. In an open chain r, exogenous arrival process at node i, $1 \leq i \leq M$, $1 \leq r \leq R$, can be load dependent or load independent. λ denotes the rate at which customers arrive to the network when the arrival rate is load independent. For load dependent arrival rate, we can distinguish the following two cases:

(i)   the rate depends on the total number of customers in the entire network and it is denoted by $\lambda\, a(n)$, $n \geq 0$;
(ii) for multichain networks, the arrival rate can depend on the total network population N_r in each chain r, denoted by $a_r(N)$ and $N = (N_1,..., N_R)$, $N_r \geq 0$, for each chain r, $1 \leq r \leq R$;

where a and a_r are arbitrary and non-negative functions.

An exogenous arrival attempts to enter node i in class s of chain r with probability $p_{0(i,s)}$, $1 \leq i \leq M$, $1 \leq s \leq C$, $s \in E_r$, $1 \leq r \leq R$. That is, the arrival process at node i for class s of chain r customers has the mean rate $\lambda p_{0(i,s)}$ for load independent arrivals, and $\lambda a(n) p_{0(i,s)}$ or $\lambda a_r(N) p_{0(i,s)}$, for load dependent arrivals. The complete definition of the arrival process includes the specification of the type of arrival rate.

A class s customer departing from node i attempts to enter node j as a class t customer with probability $p_{(i,s)(j,t)}$. The routing matrix is then defined as $P = \| p_{(i,s)(j,t)} \|$, $1 \leq i,j \leq M$, $1 \leq s,t \leq C$, $1 \leq r \leq R$. If chain r is open then $p_{(i,s)0}$, $1 \leq i \leq M$, $1 \leq s \leq C$, denotes the probability that a class s customer departs from node i upon completion of its service. By definition, the following relation holds, for $1 \leq i \leq M$, $1 \leq s \leq C$:

$$\sum_{j=1}^{M} \sum_{t=1}^{C} p_{(i,s)(j,t)} + p_{(i,s)0} = 1$$

Vector **x** with components x_{is}, $1 \leq i \leq M$, $1 \leq s \leq C$, $s \in E_r$, $1 \leq r \leq R$, can be obtained by solving the following traffic balancing system for each chain r:

$$x_{is} = \lambda \, p_{0(i,s)} + \sum_{j=1}^{M} \sum_{t \in E_r} x_{jt} \, P_{(j,t)(i,s)} \qquad (2.8)$$

No arrivals and departures take place in a closed chain r. Hence, we have $p_{0(i,s)}$ = $p_{(i,s)0}$ = 0, $\forall \ s \in E_r$. Similar to the single class case, linear system (2.8) for each closed chain r is homogeneous and it does not provide a unique solution. For open chains in queueing networks with infinite capacities, x_{is} is the throughput of class s, chain r customers at node i. It is the relative throughput or relative visit ratios of class s, chain r customers at node i if the chain is closed. This interpretation of x_{is} does not, in general, hold for queueing networks with finite capacities.

Service center i, $1 \leq i \leq M$, is described by the number of servers, the service time distribution and the service discipline. Let S_i denote the state of node i, from which one can derive vector $n_i = (n_{(i,1)}, n_{(i,2)}, \ldots, n_{(i,C)})$ with $n_{(i,s)}$ denoting the number of customers in node i in class s, $1 \leq s \leq C$. State S_i in general includes other components according to the node type that describe the service discipline and/or service time distribution and the blocking type. Let n_i and n_{ir} denote the total number of jobs at node i and in chain r, $1 \leq i \leq M$, $1 \leq r \leq R$, respectively.

The service time distribution of customers at node i, class s is denoted by $F_{is}(t)$, $t \geq 0$, $1 \leq i \leq M$, $1 \leq s \leq C$, with mean value $1/\mu_{is}$, if it is load independent. Node i service rate for jobs of class s in chain r can be also dependent on the number of customers either in node i, n_i, or in node i and chain r, n_{ir}. It is then denoted by $\mu_{is} f_i(n_i)$ and $\mu_{is} f_{ir}(n_{ir})$, respectively, n_i, $n_{ir} \geq 0$, where f_i and f_{ir} are arbitrary non-negative functions, $1 \leq i \leq M$, $s \in E_r$, $1 \leq r \leq R$.

The capacity of each node can also be chain and class dependent. Consider a multiclass queueing network with C>1 classes and R>1 chains of customers. Then B_{ir} denotes the maximum queue length at node i for chain r. This definition can also be extended to make it class dependent. Then we have $B_{(i,s)}$, $1 \leq i \leq M$, $1 \leq s \leq C$, $s \in E_r$, denoting the maximum queue length at node i for chain r, class s customers.

Finally, blocking mechanisms defined for single class networks can be extended to multiclass queueing networks. The blocking mechanism can be defined differently for each chain, for each class of customers within a chain, or it may depend on the population of either the subnetwork or the total network. Furthermore, the blocking mechanism can also define constraints on the minimum and maximum populations.

2.5   PERFORMANCE INDICES

Queueing networks are models of real life systems developed to estimate various performance metrics of the system under a set of assumptions on the system parameters. These performance metrics may be defined for each node, chain, and/or class. Most metrics of interest are defined in terms of average or mean rate of a performance measure. Some of the commonly used parameters include, for each node i:

U_i utilization
X_i throughput
L_i mean queue length
T_i mean response time

In addition, the distributions of various random variables are commonly used in detailed analysis of the system:

n_i number of customers at node i
z_i number of active servers at node i. An active server in this context refers to a server that is not empty and not blocked, $0 \leq z_i \leq \min \{n_i, K_i\}$, where K_i is the number of servers of the node
t_i customer passage time through the node.

Let $\pi_i(n_i)$ denote the stationary (marginal) queue length distribution of node i, i.e., the stationary probability of having n_i customers at node i in steady state, $n_i \geq 0$, $1 \leq i \leq M$.

Let $\xi_i(n_i)$ denote the stationary queue length distribution of node i at arrival time of a customer at that node and $\zeta_i(n_i, z_i)$ denote the stationary joint distribution of variables n_i and z_i, i.e., the probability of n_i customers and z_i active servers in node i at arbitrary times, $n_i \geq 0$, $1 \leq i \leq M$.

The definitions of these probabilities depends on the blocking type and are given in Chapter 4.

The *customer sojourn time* is the time a customer spends in the queueing network or in a part of it.

The *network sojourn time* is the time a customer spends in the queueing network and the *passage time* is the time a given customer takes to traverse a path in the network, i.e. between two consecutive departures of a customer from two given nodes.

The *node response time* is the sojourn time that refers to a customer visit to a given service center in the network. The *residence time* is the sojourn time of a customer that consists of succession of visits to a given service center in the network.

The node response time distribution is usually required in the evaluation of job residence time and passage time distributions in queueing networks.

The *cycle time* is the time between two consecutive departures of a particular customer from a node in the network.

The average sojourn, passage and cycle time can be easily derived by the node mean response time and the visit ratio. For example in a closed network the average cycle time for node i is given by $\sum\limits_{j=1}^{M} x_j T_j / x_i$ and in an open network the average

network sojourn time or network residence time is given by $\sum\limits_{j=1}^{M} x_j T_j$, where x_i's are obtained by formula (2.8). The average residence time T_i^* for node i is defined as the average node response time T_i multiplied by the visit ratio, i.e. $T_i^* = T_i x_i$.

The sojourn time distribution is an interesting and detailed performance measure of a queueing network, as well as the passage time and cycle time distribution, but few analytical results are available in the literature.

For open queueing networks with finite capacity another performance index of interest is the job loss probability, which is the probability that an arriving customer is not accepted and it is lost. This probability denoted by P_{loss} can be computed from the stationary queue length distribution at arrival times as follows for BAS, BBS and RS mechanisms:

$$P_{loss} = \sum_{1 \le i \le M} \xi_i(B_i)$$

Note that for STOP and RECIRCULATE mechanisms the arrival rate depends on the function a(n) and it becomes null when the network population n is greater or equal to the upper bound U, defined in Section 2.2.

One can derive the mean performance indices from the stationary joint distribution, which depends on the blocking type.

Consider BAS, BBS and RS blocking mechanisms. When node i cannot be blocked its utilization can be simply defined as follows:

$$U_i = \sum_{n_i=1}^{\min(B_i,N)} \frac{\min(n_i,K_i)}{K_i} \pi_i(n_i) \qquad (2.9)$$

When node i can be blocked this formula gives for each blocking type the total utilization, that is the fraction of time that the node is not empty.

It is worth noting that in queueing networks with blocking, utilization could be defined as the probability that the servers are really providing service, i.e., they are neither empty nor blocked.

Hence in networks with blocking it is also interesting to define a different performance index that we call *effective utilization* and is the fraction of time that the node is neither empty nor blocked. In other words the effective utilization is a measure of the useful work of the node. Let U_i^e denote the effective utilization of node i whose definition depends on the blocking type.

For BAS and BBS blocking, including each subtype, the effective utilization is defined by considering only the active servers as follows:

$$U_i^e = \sum_{n_i=1}^{\min(B_i,N)} \sum_{z_i=1}^{\min(n_i,K_i)} \frac{z_i}{K_i} \zeta_i(n_i,z_i)$$

For RS blocking we observe that by definition the work of node i server is useful only if at the service completion time, i.e. at departure time from node i, the destination node is not full. If node i has exponential service time distribution, due to the memoryless property one can argue that this is equivalent to consider the state of the destination node at random time, during the customer service. However, as we shall see in chapters 4 and 5, in general departure, arrival time and random time state distribution are different except for some special cases. Nevertheless we consider the state probability π to define the effective utilization for queueing networks with RS blocking as follows:

$$U_i^e = \sum_{k=1}^{\min(B_i,N)} \frac{\min(k,K_i)}{K_i} \sum_{\forall n \in E: n_i=k} \sum_{\forall j: n_j < B_j} \pi(n)p_{ij}$$

where $n = (n_1, n_2, ...,n_M)$ and E is the state space of the exponential network with RS blocking.

For single server service center, i.e., $K_i=1$, then $z_i=0,1$ and these average performance indices can be simplified as follows:

$$U_i = \sum_{n_i>0} \pi_i(n_i)$$

$$U_i^e = \sum_{n_i>0} \zeta_i(n_i,1) \quad \text{for BAS and BBS}$$

$$U_i^e = \sum_{k=1}^{\min(B_i,N)} \sum_{\forall n \in E: n_i=k} \sum_{\forall j: n_j < B_j} \pi(n)p_{ij} \quad \text{for RS}$$

For STOP blocking, since the service in each node is stopped when total network population n reaches the minimum threshold L then the effective utilization can be defined as follows:

$$U_i^e = \sum_{k=1}^{U} \frac{\min(k,K_i)}{K_i} \sum_{\forall n \in E: n_i=k,\, n>L} \pi(n)$$

where E is the state space of the exponential network with STOP blocking, L and U are the minimum and the maximum population admitted in the network, with

$0 \leq L \leq U$, and n is the total network population when the state is $\mathbf{n}=(n_1, ..., n_M)$,
$$n = \sum_{1 \leq i \leq M} n_i .$$
For RECIRCULATE blocking the nodes are never blocked, because when the network population n reaches the minimu threshold L a customer leaving node i is forced to stay in the network with probability $p_{i0}[1-d(n)]$ and moves to any node in the network according to the routing probabilities. Hence $U_i^e = U_i$.

For each blocking type node throughput is defined as follows:

$$X_i = \sum_{n_i \geq 1} \sum_{z_i \geq 1} \mu_i f_i(n_i, z_i) \zeta_i(n_i, z_i) \qquad (2.10)$$

For single server service center, i.e., $K_i=1$, then $z_i=0,1$ and the throughput can be simplified as follows:

$$X_i = \sum_{n_i > 0} \mu_i f_i(n_i) \zeta_i(n_i, 1)$$

Note that like the utilization, in networks with blocking we can also define the *effective throughput* as a measure of the useful work of the node. For BAS blocking such measure is identical to the throughput defined by formula (2.10), because the server is active only if the node is not blocked. The same holds for BBS blocking if we assume that the service interrupted by a full destination node is resumes as soon as the node becomes unblocked. On the contrary for BBS blocking if the interrupted service due to blocking is lost we observe a repetition of the work. For RS blocking, by definition a customer that tries to enter a full destination node receives a new service in the node.

Then for BBS blocking where the interrupted service is repeated and for RS blocking we can also define the *effective throughput* as a measure of the useful work of the node, given by the fraction of throughput that is not due to the service repetition because of blocking. Let X_i^e denote the effective throughput of node i for RS and BBS blocking, that similarly to the effective utilization can be defined as follows:

$$X_i^e = \sum_{k=1}^{\min(N,B_i)} \sum_{\forall \mathbf{n} \in E: n_i=k} \mu_i f_i(n_i, K_i) \sum_{\forall j: n_j < B_j} \pi(\mathbf{n}) p_{ij} \qquad (2.11)$$

where $\mathbf{n} = (n_1, n_2, ..., n_M)$ and E is the state space of the exponential network with RS blocking.

In the case of constant service rate μ_i, the node throughput and the node effective throughput for $K_i \geq 1$ servers reduces to

$$X_i = X_i^e = U_i^e K_i \mu_i$$

for BAS blocking

$$X_i = U_i K_i \mu_i$$
$$X_i^e = U_i^e K_i \mu_i$$

for RS and BBS blocking.

Mean queue length and mean response time for node i can be computed as for queueing network models with infinite capacity queues as follows:

$$L_i = \Sigma_{n_i} \ n_i \ \pi_i(n_i)$$
$$T_i = L_i \ / \ X_i \ .$$

For multichain networks, we can define the following more detailed performance indices for each node i, $1 \leq i \leq M$, and each class s, $1 \leq s \leq C$:

$U_{(i,s)}$	class s utilization at node i
$X_{(i,s)}$	class s throughput at node i
$L_{(i,s)}$	class s mean queue length at node i
$T_{(i,s)}$	class s mean response time at node i.

Note that these performance indices are defined as for the single class model.
Average performance indices for each node i, $1 \leq i \leq M$, and each chain r, $1 \leq r \leq R$, are:

U_{ir}	chain r utilization at node i
X_{ir}	chain r throughput at node i
L_{ir}	chain r mean queue length at node i
T_{ir}	chain r mean response time at node i

Performance indices related to chain r can be obtained by the indices of the classes belonging to that chain. For example, chain r utilization at node i is defined as follows:

$$U_{ir} = \sum_{s \in E_r} U_{(i,s)}$$

Similar relationships hold for throughput and mean queue length.
Chain r mean response time at node i is defined as a weighted sum of the class s mean response time, for each s at node i, as follows:

$$T_{ir} = \sum_{s \in E_r} T_{(i,s)} X_{(i,s)} / X_{i,r}$$

where the weights represent the proportion of class s jobs in respect to the overall population at node i, chain r.

For the overall network, we consider the following average performance indices:

X	throughput of the network
L	mean population in the network
T	mean residence time in the network
T(i,j)	mean passage time from node i to node j

These performance indices of the entire network can be simply derived from the node performance indices both for single class networks and for multichain networks. In particular we can write for single class networks:

$$X = X_i / x_i, \quad L = \sum_{i=1}^{M} L_i, \quad T = \sum_{i=1}^{M} x_i T_i$$

$$T(i,i) = \sum_{j=1}^{M} x_j T_j / x_i \quad \text{mean cycle time of node i}$$

$$T(i,j) = \sum_{k=1}^{M} v_k^{(i,j)} T_k \quad \text{mean passage time from node i to node j}$$

where $v_k^{(i,j)}$ is the mean number of visits to node k during the passage time of a customer from node i to node j.

Performance indices of queueing networks with finite capacity can be evaluated through the analytical or simulation techniques. The analysis of networks with blocking is presented in Part II.

2.6   BIBLIOGRAPHICAL NOTES

Queueing networks definition
The definition of queueing network models without blocking can be found in Kleinrock (1975), Baskett et al. (1975), Lavenberg (1983), that also discuss the meaning of the solution of vector x of visit ratio defined by equation (2.1).

The framework to define service disciplines given by equation (2.2) was introduced by Chandy, Howard and Towsley (1977), and further extended in Kelly (1979) and Chandy and Martin (1983). In particular Kelly (1979) presents the definitions of symmetric service discipline and Chandy and Martin (1983) discuss station balancing.

The definition of the blocking function given by formula (2.4) was introduced by Hordijk and van Dijk (1981).

Blocking mechanisms definition
The blocking mechanisms introduced in Section 2.2 are named and classified by Akyildiz and Perros (1989) and by Onvural (1990).

BAS blocking mechanism has also referred to as classical, transfer, manufacturing and production blocking in the literature by several authors, and can be found in Hillier and Boling (1967), Boxma and Konheim (1981), Neuts (1986), Onvural and Perros (1986), Suri and Diehl (1986), Akyildiz (1987), Altiok and Perros (1987), De Nitto Personè and Grillo (1987), Perros, Nilsson and Liu (1988), Akyildiz and Perros (1989), Balsamo and Donatiello (1989b), Dallery and Frein (1989), Kundu and Akyildiz (1989), Onvural (1989), Onvural and Perros (1989), Perros (1989a, 1989b), Onvural (1990), Balsamo and De Nitto Personè (1991), Balsamo and Clò (1992), Onvural (1993), Balsamo and De Nitto Personè (1994), De Nitto Personè (1994). The First Blocked First Unblocked discipline is defined in Altiok and Perros (1987) and Onvural (1990).

BBS mechanism has also referred to as service or immediate blocking in the literature, and is studied in Gordon and Newell (1967), Caseau and Pujolle (1979), Boxma and Konheim (1981), Gershwin and Berman (1981), Onvural and Perros (1986), Altiok and Perros (1987), De Nitto Personè and Grillo (1987), Gershwin (1987), Balsamo and Donatiello (1989a), Frein and Dallery (1989), Gün and Makowski (1989), Onvural (1989), Onvural and Perros (1989), Perros (1989a), Perros (1989b), Onvural (1990), Balsamo and De Nitto Personè (1991), Dallery and Towsley (1991), Balsamo and Clò (1992), (1993), Balsamo, Clò and Donatiello (1993), Onvural (1993), Balsamo and De Nitto Personè (1994) and De Nitto Personè (1994).

The classification of BBS mechanism into the two subcategories is given in Onvural (1990). BBS-O blocking was defined in Gordon and Newell (1967), Konheim and Reiser (1978), De Nitto Personè and Grillo (1987), Balsamo and De Nitto Personè (1991) and De Nitto Personè (1994).

RS blocking mechanism has also referred to as rejection, retransmission and repeat blocking in the literature, and it can be found in Kingman (1969), Caseau and Pujolle (1979), Pittel (1979), Gershwin and Berman (1981), Hordijk and van Dijk (1981), Balsamo and Iazeolla (1983), Hordijk and van Dijk (1983), Yao and Buzacott (1985), Dallery and Yao (1986), van Dijk and Tijms (1986), Onvural and Perros (1986), De Nitto Personè and Grillo (1987), Yao and Buzacott (1987), Akyildiz and von Brand (1989a), (1989b), Kouvatsos and Xenios (1989), Onvural (1989), Onvural and Perros (1989), Perros (1989a), (1989b), Akyildiz and van Dijk (1990), Onvural (1990), Balsamo and De Nitto Personè (1991), van Dijk (1991a), Onvural (1993), Balsamo and De Nitto Personè (1994) and De Nitto Personè (1994).

Generalized and Kanban blocking for tandem open networks are introduced in Buzacott (1989), Mitra and Mitrani (1990) and (1992), Cheng (1993) and in Mishra and Fang (1997).

STOP blocking, also referred to as delay blocking is introduced in van Dijk (1991a), (1991b) and (1993). Recirculate blocking, also called triggering protocol, is defined in Jackson (1963), Lam (1977) and van Dijk (1991a) and (1991b).

Kundu and Akyildiz (1989) define some algorithms for deadlock prevention based on the deadlock-free condition introduced in Section 2.2.

State-dependent routing was introduced for queueing networks without blocking in Baskett et al. (1975), Kelly (1979) and Towsley (1980). Queueing networks with state-dependent routing and blocking are considered in Hordijk and van Dijk (1981), (1983), Balsamo and Iazeolla (1986), Yao and Buzacott (1987) and Akyildiz and von Brand (1989a).

The definition of multiclass queueing networks with blocking is based on the definition of networks without blocking as presented in Kleinrock (1975) and Lavenberg (1983), as well as the definition of performance indices whose specialization for networks with blocking was introduced in Balsamo and De Nitto Personè (1991) and Perros (1994).

A survey on sojourn time distribution in queueing networks can be found in Boxma and Daduna (1990).

REFERENCES

Akyildiz, I.F. "Exact product form solution for queueing networks with blocking" IEEE Trans. on Computer, Vol. 36 (1987) 122-125.

Akyildiz, I.F., and H.G. Perros Special Issue on Queueing Networks with Finite Capacity Queues, Performance Evaluation, Vol. 10, 3 (1989).

Akyildiz, I.F., and N. Van Dijk "Exact Solution for Networks of Parallel Queues with Finite Buffers" in: Proc. *Performance '90* (P.J.B. King, I. Mitrani and R.J. Pooley Eds.) North-Holland (1990) 35-49.

Akyildiz, I.F., and H. Von Brand "Central Server Models with Multiple Job Classes, State Dependent Routing, and Rejection Blocking" IEEE Trans. on Softw. Eng., Vol. 15 (1989) 1305-1312.

Akyildiz, I.F., and H. Von Brand "Exact solutions for open, closed and mixed queueing networks with rejection blocking" J. Theor. Computer Science, Vol. 64 (1989) 203-219.

Altiok, T., and H.G. Perros "Approximate analysis of arbitrary configurations of queueing networks with blocking" Ann. Oper. Res., Vol. 9 (1987) 481-509

Altiok, T., and S.S. Stidham "A note on Transfer Line with Unreliable Machines, Random Processing Times, and Finite Buffers" IIE Trans., Vol. 14, 4 (1982) 125-127.

Balsamo, S., and C. Clò "State distribution at arrival times for closed queueing networks with blocking" Technical Report TR-35/92, Dept. of Comp. Sci., University of Pisa, 1992.

Balsamo, S., C. Clò, and L. Donatiello "Cycle Time Distribution of Cyclic Queueing Network with Blocking", in *Queueing Networks with Finite Capacities* (R.O. Onvural and I.F. Akyildiz Eds.), Elsevier, 1993, and Performance Evaluation, Vol. 17 (1993) 159-168.

Balsamo, S., and C. Clò "Delay distribution in a central server model with blocking", Technical Report TR-14/93, Dept. of Comp. Sci., University of Pisa, 1993.

Balsamo, S., and V. De Nitto Personè "Closed queueing networks with finite capacities: blocking types, product-form solution and performance indices" Performance Evaluation, Vol. 12, 4 (1991) 85-102.

Balsamo, S., and V. De Nitto Personè "A survey of Product-form Queueing Networks with Blocking and their Equivalences" Annals of Operations Research, Vol. 48 (1994) 31-61.

Balsamo, S., and L. Donatiello "On the Cycle Time Distribution in a Two-stage Queueing Network with Blocking" IEEE Transactions on Software Engineering, Vol. 13 (1989) 1206-1216.

Balsamo, S., and L. Donatiello "Two-stage Queueing Networks with Blocking: Cycle Time Distribution and Equivalence Properties", in *Modelling Techniques and Tools for Computer Performance Evaluation* (R. Puigjaner, D. Potier Eds.) Plenum Press, 1989.

Balsamo, S., and Iazeolla, G. "Some Equivalence Properties for Queueing Networks with and without Blocking", in *Performance '83* (A.K. Agrawala, S.K. Tripathi Eds.) North Holland, 1983.

Balsamo, S., and G. Iazeolla "Synthesis of Queueing Networks with Block and State-dependent Routing" Computer System Science and Engineering, Vol. 1 (1986).

Baskett, F., K.M. Chandy, R.R. Muntz, and G. Palacios "Open, closed, and mixed networks of queues with different classes of customers" J. of ACM, Vol. 22 (1975) 248-260.

Boxma, O., and A.G. Konheim "Approximate analysis of exponential queueing systems with blocking" Acta Informatica, Vol. 15 (1981) 19-66.

Boxma, O., and H. Daduna "Sojourn time distribution in queueing networks" in 'Stochastic Analysis of computer and Communication Systems' (H. Takagi Ed.) North Holland, 1990.

Buzacott, J.A. "Queueing Models of Kanban and MRP controlled production systems" Engineering Costs and Production Economics, Elsevier Science Pub., Vol. 17 (1989) 3-20.

Caseau, P., and G. Pujolle "Throughput capacity of a sequence of transfer lines with blocking due to finite waiting room" IEEE Trans. on Softw. Eng., Vol. 5 (1979) 631-642.

Chandy, K.M., J.H. Howard, and D. Towsley "Product form and local balance in queueing networks" J. ACM, Vol. 24 (1977) 250-263.

Chandy, K.M., U. Herzog, and L. Woo "Parametric analysis of queueing networks" IBM J. Res. Dev., Vol. 1 (1975) 36-42.

Chandy, K.M., and A.J. Martin "A characterization of product-form queueing networks" J. ACM, Vol.30 (1983) 286-299.

Cheng, D.W. "Analysis of a tandem queue with state dependent general blocking: a GSMP perspective" Performance Evaluation, Vol. 17 (1993) 169-173.

Choukri, T. "Exact Analysis of Multiple Job Classes and Different Types of Blocking" in *Queueing Networks with Finite Capacities* (R.O. Onvural and I.F. Akyidiz Eds.), Elsevier, 1993.

Cohen, J.W. "The multiple phase service network with generalized processor sharing" Acta Informatica, Vol. 12 (1979) 245-284.

Courtois, P.J., and P. Semal "Computable bounds for conditional steady-state probabilities in large Markov chains and queueing models" IEEE Journal on SAC, Vol. 4 (1986) 920-936.

Dallery, Y., and Y. Frein "A decomposition method for the approximate analysis of closed queueing networks with blocking", Proc. First Int. Workshop on Queueing Networks with Blocking, (H.G. Perros and T. Altiok Eds.) North Holland, 1989.

Dallery, Y., and D.F. Towsley "Symmetry property of the throughput in closed tandem queueing networks with finite buffers" Op. Res. Letters, Vol. 10 (1991) 541-547.

Dallery, Y., and D.D. Yao "Modelling a system of flexible manufacturing cells" in: *Modeling and Design of Flexible Manufacturing Systems* (Kusiak Ed.) North-Holland, 1986, 289-300.

De Nitto Personè, V., and D. Grillo "Managing Blocking in Finite Capacity Symmetrical Ring Networks" Third Int. Conf. on Data Comm. Systems and their Performance, Rio de Janeiro, Brazil, June 22-25 (1987) 225-240.

Frein, Y., and Y. Dallery "Analysis of Cyclic Queueing Networks with Finite Buffers and Blocking Before Service", *Performance Evaluation*, Vol. 10 (1989) 197-210.

Gershwin, S. "An efficient decomposition method for the approximate evaluation of tandem queues with finite storage space and blocking" Oper. Res., Vol. 35 (1987) 291-305.

Gershwin, S., and U. Berman "Analysis of transfer lines consisting of two unreliable machines with random processing times and finite storage buffers" AIIE Trans., Vol. 13 (1981) 2-11.

Gordon, W.J., and G.F. Newell "Cyclic queueing systems with restricted queues" Oper. Res., Vol. 15 (1967) 286-302.

Gün, L., and A.M. Makowski "An approximation method for general tandem queueing systems subject to blocking" Proc. First Int. Workshop on Queueing Networks with Blocking, (H.G. Perros and T. Altiok Eds.) North Holland, 1989,147-171.

Hillier, F.S., and R.W. Boling "Finite queues in series with exponential or Erlang service times - a numerical approach" Oper. Res., Vol. 15 (1967) 286-303.

Hordijk, A., and N. Van Dijk "Networks of queues with blocking", in: Performance '81 (K.J. Kylstra Ed.) North Holland (1981) 51-65.

Hordijk, A., and N. Van Dijk "Networks of queues; Part I: job-local-balance and the adjoint process; Part II: General routing and service characteristics", in: Lect. Notes in Control and Information Sciences (F. Baccelli and G. Fajolle Eds.) Springer-Verlag, 1983, 158-205.

Jackson, J.R. "Jobshop-like queueing systems" Management Science, Vol. 10 (1963) 131-142.

Kelly, F.P. *Reversibility and Stochastic Networks.* Wiley (1979).

Kingman, J.F.C. "Markovian population process" J. Appl. Prob., Vol. 6 (1969) 1-18.

Kleinrock, L. *Queueing Systems.Vol.1: Theory,* Wiley (1975).

Konhein, A.G., and M. Reiser "A queueing model with finite waiting room and blocking" SIAM J. of Comput, Vol. 7 (1978) 210-229.

Kouvatsos, D., and N.P. Xenios "Maximum entropy analysis of general queueing networks with blocking" Proc. First Int. Workshop on Queueing Networks with Blocking, (H.G. Perros and T. Altiok Eds.) North Holland, 1989.

Krzesinski, A.E. "Multiclass queueing networks with state-dependent routing" Performance Evaluation, Vol.7 (1987) 125-145.

Kundu, S., and I. Akyildiz "Deadlock free buffer allocation in closed queueing networks" Queueing Systems Journal, Vol. 4 (1989) 47-56.

Lam, S.S. "Queueing networks with capacity constraints" IBM J. Res. Develop., Vol. 21 (1977) 370-378.

Lavenberg, S.S. *Computer Performance Modeling Handbook.* Prentice Hall, 1983.

Lavenberg, S.S., and M. Reiser "Stationary State Probabilities at Arrival Instants for Closed Queueing Networks with multiple Types of Customers" J. Appl. Prob., Vol. 17 (1980) 1048-1061.

Mishra, S., and S.C. Fang "A maximum entropy optimization approach to tandem queues with generalized blocking" Performance Evaluation, Vol. 30 (1997) 217-241.

Mitra, D., and I. Mitrani " Analysis of a Kanban discipline for cell coordination in production lines I" Management Science, Vol. 36 (1990) 1548-1566.

Mitra, D., and I. Mitrani " Analysis of a Kanban discipline for cell coordination in production lines II: Stochastic demands" Operations Research, Vol. 36 (1992) 807-823.

Neuts, M.F. "Two queues in series with a finite intermediate waiting room" J. Appl. Prob., 5 (1986) 123-142.

Onvural, R.O. "A Note on the Product Form Solutions of Multiclass Closed Queueing Networks with Blocking" Performance Evaluation, Vol.10 (1989) 247-253.

Onvural, R.O. "Survey of Closed Queueing Networks with Blocking" ACM Computing Surveys, Vol. 22 (1990) 83-121.

Onvural, R.O. Special Issue on Queueing Networks with Finite Capacity, Performance Evaluation, Vol. 17 (1993).

Onvural, R.O., and H.G. Perros "On equivalences of blocking mechanisms in queueing networks with blocking" Oper. Res. Letters, Vol. 5 (1986) 293-297.

Onvural, R.O., and H.G. Perros "Some equivalencies on closed exponential queueing networks with blocking" Performance Evaluation, Vol.9 (1989) 111-118.

Perros, H.G. "Open queueing networks with blocking" in *Stochastic Analysis of Computer and Communications Systems* (Takagi Ed.), North Holland, 1989.

Perros, H.G. "A bibliography of papers on queueing networks with finite capacity queues" Performance Evaluation, Vol. 10 (1989) 225-260.

Perros, H.G. *Queueing networks with blocking.* Oxford University Press, 1994.

Perros, H.G., A. Nilsson, and Y.G. Liu "Approximate analysis of product form type queueing networks with blocking and deadlock" Performance Evaluation, Vol. 8 (1988) 19-39.

Perros, H.G., and P.M. Snyder "A computationally efficient approximation algorithm for analyzing queueing networks with blocking" Performance Evaluation, Vol. 9 (1988/89) 217-224.

Pittel, B. "Closed exponential networks of queues with saturation: the Jackson-type stationary distribution and its asymptotic analysis" Math. Oper. Res., Vol. 4 (1979) 367-378.

Sevcik, K.S., and I. Mitrani "The Distribution of Queueing Network States at Input and Output Instants" J. of ACM, Vol. 28 (1981) 358-371.

Shantikumar, G.J., and D.D. Yao "Monotonicity properties in cyclic networks with finite buffers" Proc. First Int. Workshop on Queueing Networks with Blocking, (H.G. Perros and T. Altiok Eds.) North Holland, 1989.

Suri, R., and G.W. Diehl "A variable buffer size model and its use in analytical closed queueing networks with blocking" Management Sci. Vol.32 (1986) 206-225.

Towsley, D.F. "Queueing network models with state-dependent routing" J. ACM, Vol. 27 (1980) 323-337.

Van Dijk, N. "On 'stop = repeat' servicing for non-exponential queueing networks with blocking" J. Appl. Prob., Vol. 28 (1991) 159-173.

Van Dijk, N. "'Stop = recirculate' for exponential product form queueing networks with departure blocking" Oper. Res. Lett., Vol. 10 (1991) 343-351.

Van Dijk, N. "On the Arrival Theorem for communication networks" Computer Networks and ISDN Systems, Vol. 25 (1993) 1135-1142.

Van Dijk, N.M., and H.C. Tijms "Insensitivity in Two Node Blocking Models with Applications" in *Teletraffic Analysis and Computer Performance Evaluation* Boxma, (Cohen and Tijms Eds.), Elsevier Science Publishers, North Holland, 1986, 329-340.Whittle, P. "Partial balance and insensitivity" J. Appl. Prob. 22 (1985) 168-175.

Yao, D.D., and J.A. Buzacott "Modeling a class of state-dependent routing in flexible manufacturing systems" Annals of Oper. Res., Vol. 3 (1985) 153-167.

Yao, D.D., and J.A. Buzacott "Modeling a class of flexible manufacturing systems with reversible routing" Oper. Res., Vol. 35 (1987) 87-93.

3 APPLICATION EXAMPLES OF QUEUEING NETWORKS WITH BLOCKING

As discussed in chapter 1, customers in real systems usually require different services provided by different servers. During this process, customers may wait in different queues in front of servers prior to start receiving services. Such complex service systems are often modeled using a network of queueing systems, referred to as a *queueing network*. The topology of the network represents the flow of customers from one service station to another to meet their service requirements. Hence, a queueing network is a connected directed graph whose nodes represent the service centers. The waiting area in front of a server is represented as a queue. That is, each service station has a queue associated with it. A connection between two servers indicates the one-step moves that customers may make from one service center to another service center. The route that a customer takes through the network may be deterministic or random. Customers may be of different types and may follow different routes through the network.

As introduced in chapter 1, queueing networks are classified as open, closed or mixed. In an open model, customers enter the network from outside, receive service at one or more nodes, and eventually leave the network. Figure 3.1 illustrates an open network.

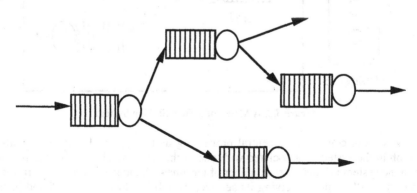

Figure 3.1. An Open Network

To model a system using an open network assumes that the arrivals to the system occurs from an infinite population of customers. That is, the rate at which customers arrive to the network is independent of the number of customers already in the system. If the user population is finite, then those already in the system are no longer candidates for entering to the network. In reality, as the number of customers in the network increases the available population dwindles, and the arrival rate falls off because of the reduced population that can generate arrivals to the network. In particular, most real life systems have finite input populations. For example, in time sharing systems, the number of jobs is limited by the number of terminals and the total number of jobs in the system is bounded. In multiprogramming systems, the degree of multiprogramming is limited by the memory size. Similarly, in communication networks, the number of unacknowledged packets in a region of the network is limited by the window size. Open networks may not be used to model such systems in which the nature of the arrival process depends strongly on the number of customers already in the system.

In closed queueing networks, there is a fixed population of customers circulating in the network. That is, no arrivals to or departures from the network are allowed. Modeling real life systems as closed queueing networks is based on the assumption that the number of customers in the system is bounded. As an example, consider the simplistic view of a time sharing system illustrated in figure 3.2.

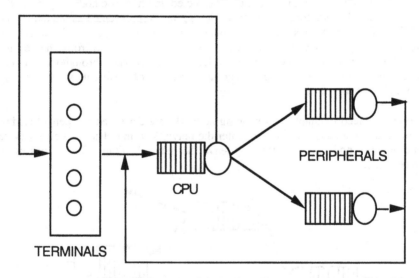

Figure 3.2. A Model of a Time Sharing System

The system consists of a central processing unit (CPU) and two peripherals. Each job in the system is associated with one of the terminals, hence the number of jobs in the system is equal to the number of terminals. A job generated by a terminal goes to the CPU. Upon receiving its service, the job is either completed and goes back to the terminals node or requests an input/output operation and joins one of the peripheral devices node with respective probabilities. Upon completion of its service at the peripheral device, the job always goes back to the CPU node.

Closed queueing networks are also used to model service systems in which the number of customers in the system is constant for a long period of time, and there is always a customer waiting to enter to as soon as a departure occurs from the system. In this case, there is always fixed number of customers circulating in the network.

As an example, consider the simplified view of a packet switched network with fixed routing as illustrated in figure 3.3. A physical network path is set up for each user session and is released when the session is terminated. End-to-end flow control is exercised to prevent buffer congestion at the exit node due to the fact that remote sources are sending traffic at a higher rate than can be accepted by the hosts fed by the exit node. Sliding window strategies are among the most popular forms of end-to-end flow control. In this scheme, the number of packets that can be outstanding without an acknowledgement between any source-destination pair is constrained to be no more than some positive integer w, called the window size.

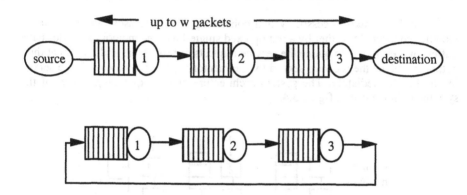

Figure 3.3. A network path with a window flow control and equivalent closed queueing model

If the network is operating at a high load condition, the source can be assumed to have a backlog of packets ready to send into the network as long as (and as soon as) the window size allows. In this case, when a packet is delivered to the destination, a new packet immediately enters the network. Hence, the system can be modeled as a closed network with w customers in it.

Often, queueing networks are analyzed with infinite queues. In an infinite queue there is always a space in the queue for arriving customers. This assumption allows queueing networks to be analyzed efficiently, under a variety of assumptions on the service distributions and service disciplines.

In real life systems, the space in front of a service station is always finite. For example, the number of cars that can wait between a gas pump and car wash station is always restricted. The number of parts that can be stored temporarily between two machines in an assembly line is always limited. The number of jobs that can wait for service by a CPU in a computer is always bounded by the amount of memory space. Hence, a more realistic model of real service systems may require queueing models with finite queue capacities to reflect finite resources associated with each server.

An important feature of queueing networks with finite queues is that the flow of customers through a node may be momentarily stopped when another node in the network reaches its capacity. That is *blocking* occurs.

For example, if the space between a gas pump and car wash station is occupied by cars, the driver that just complete purchasing may not be able to move a way from the pump, if he is waiting in the queue for a car wash. This in turn blocks the gas pump and the driver waiting behind the blocked car may not start filling the gas. As another example, consider a simplistic view of a computer communications system. The individual queues represent the finite space that is available for intermediate storage and servers correspond to communication channels. A message may not be transmitted until the destination node has space available to store the message, thus, sometimes causing the blocking of communication to that node.

In production systems, intermediate storage areas have finite capacities. A unit completing its service at a station may be forced to occupy the machine until there is a space available in the next station. While the unit blocks the machine, it may not be possible for the machine to process other units waiting in its queue.

Consider a multiprocessor system consisting of N processors and M memory modules connected together by a multiplexed single-bus. The memory modules have buffers at their inputs to queue the service requests of processors and buffers at their outputs to queue the requests served by the memory modules that can not be served by the bus immediately. The system architecture and the queueing model of the system are illustrated in figure 3.4.

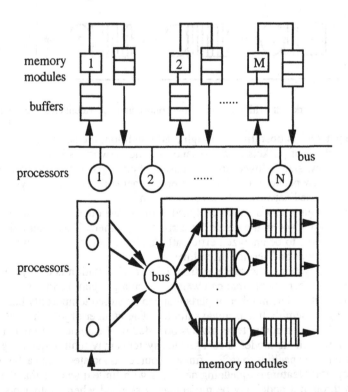

Figure 3.4. A multiprocessor architecture and its queueing model

Processor i makes a request to memory module j. If the bus at that moment is not busy transferring a request for another processor or data from a memory module, processor i takes the bus and the request is sent to memory module j. If the bus is busy transferring data, then processor i has to retry its request at a later time. If the memory module j is free it will serve the request. If it is not free, the request is queued. After the memory module completes its service, the output is placed in its output buffer for the bus to transmit it to the processor that made the request. The effect of a full node on its upstream nodes (nodes that have a directed arc to the full node) depends on the type of system being modeled. If the input buffers of the memory modules are full, then the bus can not place the request to the buffer, and the processor has to send a new request. The request will be transmitted number of times until it is delivered by the bus at a moment that there is a space at the buffer. Similarly, the output buffer of a memory module can be full. In this case, the module may be forced to suspend its service until a request is delivered from its output buffer to the processor that made the request, i.e. until a space becomes available at the output buffer. Hence, distinct models for blocking have been reported in the literature to model various real life systems with finite resources.

For presentation purposes, consider the simplistic view of a transaction processing (TP) system as shown in figure 3.5.

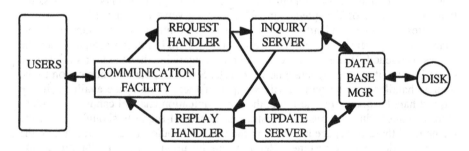

Figure 3.5. A Transaction Processing System

Users of the system send their requests (read or update) to a request handler. When the request handler receives a request from a user, it evaluates the request and passes it to an appropriate server that is designed to handle that request type. Figure 3.5, being very simplistic, illustrates two types of servers: one for handling inquiries and one for handling updates. The servers usually interact with the data base manager to gain access to (or to update) data in the database. It then formulates a reply and returns it to the user that made the request via a reply handler.

A queueing model of this transaction processing system is shown in figure 3.6. User requests are transmitted from user terminals to request handler (REQ) over a communication channel (CC). Once the request is received and processed by the request handler, it is passed to the queue of the appropriate server: inquiry server (INQ) or update server (UP). The server processes the request and sends it to the data base manager (DBM) which accesses the disk to serve the request. Once the request is served by the disk, the data (for inquiry operation) or status (for update operation) is returned to the appropriate server.

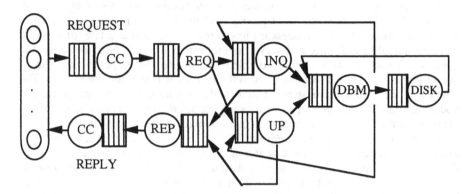

Figure 3.6. A Queueing model of a TP system

When the server completes its processing, it sends a reply to the reply handler (REP) which in turn returns the reply to the user via the communication link. The queues between various servers represent the buffers available for intermediate storage. Since there are finite number of buffers available at each node, it is possible that one or more of the queues are full at any given time. In particular, a user generates a request and attempts to access the communication channel. If the channel is busy transferring another request and if there is a buffer available then the request is queued. However, if there is no buffer available then the user suspends its operation, i.e. it can not generate new requests. Similarly, the communication to the request handler may be temporarily stopped if there is no space available in the request handler queue. Furthermore, the communication channel can not be used to store a request due to physical constraints, hence all requests should wait in the queue until there is a space available in the request handler queue at which time the communication may be resumed. Request handler upon completing the processing of a request attempts to place the request at appropriate server's queue. If there is no space available at the destination node, the request handler suspends its service until a space becomes available at the destination node. Finally, the data base manager may have to send the request to the disk a number of times before a space becomes available at the disk buffers. Hence, the effect of a full node on the service of the nodes that are connected to it may be quite different depending on the type of system being modeled. In particular, a full node may not affect the operation of an upstream server until the service is completed (user request sent to the communication channel) or the server may not start serving its customer until there is a space available in the destination node (communication channel may not transmit if there is no space available in the request handler queue). Service may be suspended during the period that the destination node is full or a customer may be served a number of times until it is accepted by the destination node (data base manager sends a request a number of times until there is a space available in the disk buffers).

A disk to tape back up model illustrated in figure 3.7 comprises of three servers and two finite buffers between servers. The first server is the disk and channel that transfers blocks of data from the disk to the main memory. The second server, the

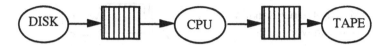

Figure 3.7. Disk to Tape Backup System

Central Processing Unit (CPU), transfers data from the main memory to the tape drive. The last server represents the tape drive. One of the performance objectives of interest is the tape back up rate (i.e. the throughput of the system). Blocking occurs due to finite spaces available for intermediate storage.

The next example is motivated by a simple Mass Storage System (MSS) as might be used in a data processing environment. The system consists of a first MSS, a staging disk, a CPU, an outstaying disk, and another MSS as illustrated in figure 3.8. Due to the relatively small sizes of the buffers, the blocking primarily occurs between the disks and the MSS devices.

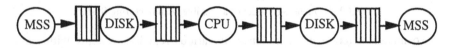

Figure 3.8. Mass Storage System

A simple terminal concentrator consists of a number of terminals, a concentrator, and a channel to transfer data to the main memory. The system configuration is the same as the disk to tape backup system illustrated above in figure 3.7 with the concentrator, the channel and the CPU replacing the disk, the CPU and the tape respectively. Similar to the above two examples, the two buffers in this terminal concentrator system have finite capacities that cause blocking of respective nodes. We note that the above examples are only sub-systems of larger configurations of computer systems, used only to illustrate the possibility of blocking due to finite storage capacities between the devices of such systems. For example, consider the disk to tape backup system. The first server corresponds to both the disk and the channel. If there is no space available in the memory, then the transfer of data has to be suspended. The server will resume its operation when a space becomes available at the memory. Similarly, other servers are forced to suspend their services if there is no space available at their destination nodes.

Similarly, consider a manufacturing system consisting of a network of automated work stations (WS) linked by a computer controlled by a material handling device (MHD) to transport work-pieces that are to be processed from one station to another as illustrated in figure 3.9.
    In these systems, if a work-piece finds the next station full, then it has to wait for the next turn of the MHD. At the next turn, there are two possibilities:
    i) the work-piece can only be processed by one station, therefore, the next attempt can only be made to the previously chosen station, or
    ii) the unit may be processed by all stations, hence, the next station is chosen independent of the previous choice(s).

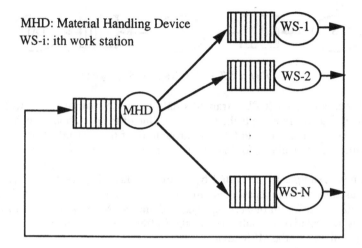

Figure 3.9. A Queueing Model of a Flexible Manufacturing System

To model different characteristics of various real life systems with finite resources, various blocking mechanisms that define distinct models of blocking have been reported in the literature. In particular, each blocking mechanism defines when a node is blocked, what happens during the blocking period, and how a node becomes unblocked, as defined formally in chapter 2.

In the examples above the different behaviors of the servers correspond to different blocking mechnisms. In particular:

• the production system described above can be represented by a queueing network model with BAS blocking, because when a unit completes its service it is forced to occupy the machine until there is a space available in the next station and the machine is blocked;

• in the multiprocessor architecture described in figure 3.4, if the bus is busy then processor i retries its request at a later time and this can be modeled by Repetitive Service mechanism. Differently, the behavior of a memory module can be modeled as Blocking Before Service, because if the output buffer of a memory module is full then the module suspends its service until a request is delivered from the buffer, that is until a space becomes available at the output buffer;

• in the queueing model of a TP system shown in figure 3.6, the behavior of a user when the channel is busy can be represented by BBS-SO. The same mechanism can be used to model the behavior of the request handler, since it suspends its service until a space becomes available at the destination node. Server CC when the request handler queue is full corresponds to BBS-SNO mechanism since the communication channel can not be used to store a request. Finally, the behavior of the data base manager is modeled by RS blocking since it may have to send the request to the disk a number of times before a space becomes available at the disk buffers;

• in the last example for the Flexible Manufacturing System illustrated in figure 3.9, the processing of a work-piece by the MHD may be modeled by a service center with RS-FD or RS-RD blocking respectively for the two cases i) and ii) described above.

Queueing networks are used to analyze the performance characteristics of flow based systems which covers pretty much any system where entities arrive at the system, are processed at one or more stations, and leave the system after its processing requirements are fullfilled. The accuracy of the results of these models in predicting the performance of the system depends on the underlying assumptions such as arrival time distributions, service time distributions, scheduling disciplines, and queue capacities. In some cases, assuming that each queue has an infinite capacity may be acceptable if the probability of each queue being full is negligible (i.e., there is a very high probability that an arriving entity finds an available space to join the waiting area of the station). Often, however, it is necessary to impose limitations on the queue sizes and analyze the system with a queueing network with finite capacity queues.

In this chapter, we have presented a small number of examples of real systems with finite queue capacities. When a queue reaches its capacity, service at its upstream stations are blocked. What happens when a server is blocked depends on the system modeled. Different examples presented in this chapter motivates different blocking mechanisms defined in the literature.

In the next chapter we present the formal definition of the Markov process underlying the queueing network model for each blocking type defined in chapter 2.

REFERENCES

Highleyman, W. H. *Performance Analysis of Transaction Processing Systems.* Prentice Hall, Inc., Englewood Cliffs, New Jersey, 1989.

Kleinrock, L. *Queueing Systems. Vol.1: Theory.* Wiley, 1975.

Lavenberg, S.S. *Computer Performance Modeling Handbook.* Prentice Hall, 1983.

Onvural, R.O. Survey of Closed Queueing Networks with Blocking. ACM Computing Surveys, Vol. 22, 2 (1990) 83-121.

Onvural, R.O., and Perros, H.G. "On Equivalencies of Blocking Mechanisms in Queueing Networks with Blocking" Oper. Res. Letters, Vol. 5 (1986) 293-298.

Reiser, M. "A Queueing Network Analysis of Computer Communications Networks with Window Flow Control" IEEE Trans. on Comm., Vol. 27 (1979) 1199-1209.

Yao, D.D., and Buzacott, J.A. "Modeling a Class of State Dependent Routing in Flexible Manufacturing Systems" Annals of Operations Research, Vol. 3 (1985) 153-167.

Yao, D.D., and Buzacott, J.A. "Queueing Models for Flexible Machining Stations Part II: The Method of Coxian Phases" Eur. J. Operations Research, Vol. 19 (1985) 241-252.

Yao, D.D., and Buzacott, J.A. "The Exponentialization Approach to Flexible Manufacturing System Models with General Processing Times" Eur. J. of Operations Research, Vol. 24 (1986) 410-416.

PART II

ANALYSIS OF QUEUEING
NETWORKS WITH BLOCKING

4 EXACT ANALYSIS OF MARKOVIAN NETWORKS

In this chapter we deal with analytical solutions of a class of networks that can be represented by a continuous-time Markov process.

The exact analysis of queueing networks with finite capacity and blocking concerns the evaluation of a set of performance measures that includes:
1) mean performance indices and joint queue length distribution at arbitrary times
2) stationary joint queue length distribution at arrival times
3) passage time and cycle time distributions.

In Section 4.1, we introduce the Markov process associated to queueing networks with blocking for different blocking mechanism and present the solution to evaluate the queue length distribution and average performance indices in Section 4.2. In Sections 4.3 and 4.4 we describe the processes to evaluate the arrival time queue length and the passage time distribution, respectively.

4.1   THE MARKOV PROCESS AS THE NETWORK MODEL

In this section, we introduce the definition of the Markov process associated with queueing networks with finite capacity and blocking. This detailed definition allows us to define an exact solution method to evaluate the joint queue length distribution of nodes in steady-state conditions and to evaluate average performance indices.

External arrivals are assumed to be Poisson whereas the service time distribution of each service center is defined using the class of phase-type distributions.

We assume independence between successive services of the nodes and the interarrival times. Under these assumptions, the queueing network can be represented by a continuous-time homogeneous Markov chain. The stationary and transient behaviour of the network can be analyzed by this underlying Markov process.

We consider these systems in stationary conditions. That is, we are interested in the analysis of systems with finite capacity resources without deadlock that reach a steady-state behavior thereby allowing us to evaluate the performance indices of the modelled system in equilibrium.

Under these assuptions the queueing network model reaches a steady-state condition and the underlying Markov chain has a stationary state distribution. This condition holds when the process is ergodic, that is irreducible, positive recurrent and aperiodic. Moreover, if a process has a finite state space, the process irreducibility condition is sufficient to have stationary state distribution. The underlying Markov process of a queueing network with finite capacity queues has finite state space if the network is closed or if it is open and there is an upper bound on the network population or each queue has finite capacity. In these cases the Markov process is irreducible when the routing matrix of the network is irreducible.

Next we shall consider deadlock-free queueing networks in steady-state conditions.

The definition of the system state and the underlying Markov process of the queueing network with finite capacity and blocking depends both on network characteristics and on the blocking type. We first define system state for the single class network model with blocking introduced in Chapter 2.

For the sake of simplicity we consider load independent arrivals, exponential service time distribution and the First Come First Served discipline. The extension to more general service time distributions and policies leads to the introduction of additional components to system states in a similar way as in queueing networks with infinite capacity queues. For this reason in order to define such additional state components that only depend on the node type and are independent of the blocking mechanism, it is sufficient to refer to state definitions introduced for networks with infinite capacity. For example, for queueing networks with BCMP-type nodes one can refer to the state definition of the product-form BCMP networks to complete the state definition of queueing networks with blocking defined below.

Let $S=(S_1,...,S_M)$ denote the system state at random time, where S_i represents the node i state ($1 \leq i \leq M$), whose definition depends on the type of blocking. State S_i always includes the number of customers in node i, n_i, $1 \leq i \leq M$. If node i has finite capacity queue then $n_i \leq B_i$, $1 \leq i \leq M$.

Component n_i must satisfy the following constraints for an open network:

$$0 \leq n_i \leq B_i \qquad (4.1)$$

For a closed network with N customers:

$$a_i \leq n_i \leq B_i \qquad (4.2)$$

and

$$\sum_{i=1}^{M} n_i = N \qquad (4.3)$$

where

$$a_i = \max\left\{ 0, N - \sum_{j=1, j\neq i}^{M} B_j \right\}$$

is the minimum possible population in node i.

Let $\mathbb{B}$ = {BAS, BBS-SO, BBS-SNO, BBS-O, RS-RD, RS-FD, Stop, Recirculate} denote the set of blocking types introduced in Chapter 2.

The continuous-time Markov process associated with the network and blocking type X, $X \in \mathbb{B}$, has state space denoted by E_X. Let $\mathbf{Q_X}$ denote the infintesimal generator or transition rate matrix of the process. Each process state transition corresponds to a particular set of events on the network model, such as a job service completion at a node and the simultaneous transition towards another node or an external arrival at a node. This correspondence depends on the blocking type as described next.

We assume that there exists a stationary state distribution of the Markov process associated to the network model. According to the above definitions, if the routing matrix P is irreducible and the process state space is finite, the Markov process is irreducible on space E_X, and there exists the stationary state distribution π_X= {$\pi_X(S)$, $S \in E_X$}, which can be obtained as the unique solution of the linear system of the global balance equations

$$\pi_X \mathbf{Q_X} = 0 \qquad\qquad (4.4)$$

subject to the normalising condition $\sum_{S \in E_X} \pi_X(S)=1$, where $\mathbf{0}$ is the all zero vector.

The definition of state space E_X and transition rate matrix $\mathbf{Q_X}$ depends on the network characteristics and on the blocking type of each node. In the following sections we define the continuous-time Markov process for each blocking type: RS blocking types in Sections 4.1.1 and 4.1.2, BAS blocking in Section 4.1.3, BBS blocking types in Sections 4.1.4, 4.1.5 and 4.1.6. Section 4.1.7 provides an example of queueing network with various blocking mechanisms and the associated Markov process for each case. The Stop and Recirculate blocking are studied in Section 4.1.8. Finally, the more complex definition of the Markov process for heterogeneous networks is given in Section 4.1.9.

4.1.1 Repetitive Service Blocking - Random Destination

By definition servers in the network with RS-RD blocking cannot be blocked. In other words, the server is always active and servicing a customer if $n_i>0$. Therefore node i state definition is simply $S_i = n_i$.

For a closed network with N customers, the state space E_{RS-RD} is defined as follows:

$$\text{ERS-RD} = \{(n_1, n_2,..., n_M) \mid a_i \leq n_i \leq B_i, \ 1 \leq i \leq M, \ \sum_{i=1}^{M} n_i = N \}$$

and for an open network it is defined as follows:

$$\text{ERS-RD} = \{(n_1, n_2,..., n_M) \mid 0 \leq n_i \leq B_i, \ 1 \leq i \leq M\}.$$

For this blocking mechanism, each process state transition corresponds to one of the following cases:

- an external arrival at a given node
- a departure from the network from a given node
- a transition between pair of nodes
- a loop transition on a node due to a full destination node, i.e. blocking.

The last state transition corresponds to the blocking phenomenon. However, we observe that the server activity is never stopped. For example when blocking occurs because node i is full, jobs cannot move from a node j to node i. By repetitive service mechanism however, jobs can move from j to other non-full nodes. Hence when there is a full destination node we can still observe job transitions from a sending node to other destination nodes, corresponding to process state transitions. If all destination nodes of node j are full, the job loops back to node j and this corresponds to the process loop transition on the system state. Eventually at least one destination node of node j becomes non-full (by deadlock-free assumption this event has non-zero probability) so resuming jobs transitions from node j towards other destinations.

In order to define state transitions we have to take into account the possibility of an empty node and of a full node. Let

$$\delta(n_i) = \begin{cases} 0 & \text{if } n_i = 0 \\ 1 & \text{if } n_i > 0 \end{cases} \quad \text{and} \quad b_i(n_i) = \begin{cases} 0 & \text{if } n_i = B_i \\ 1 & \text{if } n_i < B_i \end{cases} \quad (4.5)$$

Furthermore, let $\mu_j \ f_j(n_j, K_j)$ denote the job service rate of node j with K_j servers and n_j jobs in it and e_i denote the M-vector with all zero components except one in i-th position.

Table 4.1 shows the non-zero transition rates of the infinitesimal generator matrix $\mathbf{Q}_{\text{RS-RD}} = \|q(S,S')\|$. We denote by $q(S,S')$ the transition rate between states $S,S' \in \text{ERS-RD}$. The first three rows correspond to the three transition types introduced above and the last row defines the diagonal elements of $\mathbf{Q}_{\text{RS-RD}}$.

In particular the first row corresponds to an external job arrival at node j. The arriving job can be accepted only if $n_j < B_j$ and this is considered by the blocking function $b_j(n_j)$. This transition causes the state of node j, S'_j to change to $n'_j = n_j + 1$.

The second row corresponds to a job departure from node j. A departure is possible only if $n_j > 0$ as indicated by $\delta(n_j)$. This transition causes the state S'_j to change to $n'_j = n_j - 1$.

Table 4.1. Generator matrix $Q_{RS\text{-}RD}$

$q(S,S')$	relation between S and S'
$\lambda\, p_{0j}\, b_j(n_j)$	$S' = S + e_j$
$\delta(n_j)\, \mu_j\, f_j(n_j, K_j)\, p_{j0}$	$S' = S - e_j$
$\delta(n_j)\, \mu_j\, f_j(n_j, K_j)\, p_{ji}\, b_i(n_i)$	$S' = S + e_i - e_j,\ i \neq j$
$-\displaystyle\sum_{S'' \in E_{RS-RD}, S'' \neq S} q(S,S'')$	$S' = S$

The third row corresponds to a job transition from node j towards node i. The transition is possible only if node j is not empty and node i is not full, conditions which are expressed by functions $\delta(n_j)$ and $b_i(n_i)$. The transition leads to changes in node j and i populations, in particular $n'_j = n_j - 1$ and $n'_i = n_i + 1$.

Finally, the diagonal entry of the generator matrix $Q_{RS\text{-}RD}$ defined in the last row includes the loop transitions due to a job attempt to enter a full destination node.

4.1.2 Repetitive Service Blocking - Fixed Destination

As illustrated in Chapter 7, RS-FD blocking is equivalent to BBS-SO blocking for exponential queueing networks in which when a destination is full the jobs loop back according to LIFO discipline (that is they immediately receive a new service). In terms of policies behavior, the difference between the two mechanisms is that the destination of the job is known after (RS-FD) or before (BBS-SO) the service. Moreover, when the destination node is full, in a BBS-SO network the service is blocked, while in RS-FD, the server continues to serve the job destined to the full node.

However, because of the exponential assumption and the independence between routing and service, the Markov chains associated to RS-FD and BBS-SO blocking types are identical both as state and state transitions definitions.

In particular according to RS-FD mechanism, when a customer completes the service at node i and tries to enter a saturated destination node, it does not change its destination. Hence the information on the destination node of the next customer that will exit from node i has to be included in state S_i.

We defer the RS-FD process definition to section 4.1.4 that deals with the equivalent BBS-SO mechanism.

4.1.3 Blocking After Service

For BAS blocking we have to consider (i) the server activity and (ii) the scheduling of the nodes that are blocked by a full destination node. The process state of node i can be defined as follows:

$$S_i = (n_i, s_i, \mathbf{m}_i)$$

where:
- n_i is the number of jobs in node i;
- s_i, with $0 \leq s_i \leq \min\{n_i, K_i\}$, is the number of servers of node i blocked by a full destination node and therefore containing a served job;
- $\mathbf{m}_i$ is the list of nodes blocked by node i defined as follows:

$$\mathbf{m}_i = \begin{cases} \varnothing & \text{if } n_i \leq B_i \\ \left[r_1, r_2, \ldots, r_{\tau(i)} \right] & \text{if } n_i = B_i \end{cases}$$

where r_j is the j-th node blocked by node i, for $1 \leq j \leq \tau(i)$;
$\tau(i)$ is the number of servers currently blocked by node i and

$$0 \leq \tau(i) \leq \sum_{t \in Send_i} K_t,$$

where K_t is the number of servers of node t and with
$Send_i = \{h | p_{hi} > 0, \ 1 \leq h \leq M\}$ denotes the set of node i senders.

Note that when node i is full, i.e. $n_i = B_i$, list $\mathbf{m}_i$ is empty if no node has attempted to send a job to node i. When $\mathbf{m}_i$ is not empty, it contains the indices of the nodes that have attempted to send a job to node i and are still blocked by node i. Note that $\mathbf{m}_i$ may contain more than one occurrence of the same node $h \in Send_i$ if more servers of node h are blocked by node i.

The "unblocking" scheduling is represented by list $\mathbf{m}_i$ whose node indices are ordered according to the time at which they will be unblocked. For example, for the First Blocked First Unblocked (FBFU) discipline $\mathbf{m}_i$ is a queue data structure.

The complete state notation $S_i = (n_i, s_i, \mathbf{m}_i)$ may be simplified when some components are not necessary. For example, if node i cannot be saturated ($B_i \geq N$) component $\mathbf{m}_i$ is useless because no node may be blocked waiting for room in i. If all the destination nodes of node i have infinite capacity queues, component s_i is useless because node i cannot be blocked.

For a closed network with N customers, the process state space E_{BAS} is defined as follows:

$$E_{BAS} = \{(S_1, S_2, ..., S_M) \mid S_i = (n_i, s_i, m_i) \; a_i \le n_i \le B_i, \; 1 \le i \le M, \; \sum_{i=1}^{M} n_i = N \}$$

where $S_i = (n_i, s_i, m_i)$ is defined above.
For open networks:

$$E_{BAS} = \{(S_1, S_2, ..., S_M) \mid S_i = (n_i, s_i, m_i), 0 \le n_i \le B_i, \; 1 \le i \le M\}.$$

For this blocking mechanism, differently from RS mechanism, the process state transitions do not necessarily correspond to a job transition (i.e. arrival or departure). Indeed, some process state transitions correspond to the occurrence of the blocking phenomenon and do not correspond to customer movements in the network. Moreover, according to BAS definition by effect of simultaneous transitions, when a full node ends a service and in one of its upstream node there is a served job waiting, two simultaneous transitions take place: the transition from the full node and the transition from its upstream node. Multiple transitions can take place because of a chain of blocked nodes. In the following we show some examples of simultaneous transitions.

Example 4.1. Figure 4.1 shows a central server network with M=4 nodes. Assume that only node 1 has finite capacity with $B_1=5$, while $B_2=B_3=B_4= \infty$. Each node has a single server ($K_i=1 \; \forall i$) and there are N=11 customers in the network. State notation of each

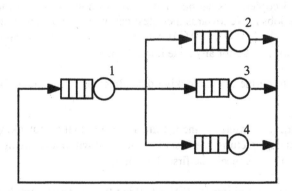

Figure 4.1. A central server network

node i=2,3,4 can be shortened as $S_i = (n_i, s_i)$, because these nodes cannot become full and as a consequence the component m_i is not needed.

Consider the network state $S=((5,0,\varnothing), (2,0), (2,0), (2,0))$ where node 1 is full but no node is blocked on it, that is $m_1=\varnothing$ and $s_i=0$ for i=2, 3, 4. As soon as node 2 completes the service, blocking occurs. That is the server of node 2 becomes blocked ($s_2=1$) and the job in it starts to wait for room in node 1 ($m_1=[2]$). This transition causes the state to change to $((5,0,[2]), (2,1), (2,0), (2,0))$.

Now, consider state ((5,0,[2, 4, 3]), (2,1), (2,1), (2,1)) where node 1 is full and all its upstream nodes are blocked with a served job waiting for node 1. List $\mathbf{m}_1$ is ordered according to the unblocking discipline and node 2 is the first blocked node. If node 1 ends the service towards node 3, two simultaneous job transitions take place: the job transition from 1 to 3 and the transition of the already served job from 2 to 1. The process transition is to state ((5,0,[4, 3]), (1,0), (3,1), (2,1)) where node 2 is now unblocked and list $\mathbf{m}_1$ is updated.

Example 4.2. Consider a closed cyclic network as shown in Figure 4.2 with M=4 nodes having finite capacity queue B_i=2 and single servers (K_i=1, $\forall i$), and N=7 customers.

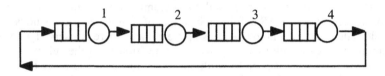

Figure 4.2. A closed cyclic network

Consider the system state ((2,1,[4]), (2,1,[1]), (2,0,[2]), (1,1,∅)) where nodes 1, 2 and 4 are blocked (s_1=s_2=s_4=1) with jobs waiting the full nodes 2, 3 and 1 ($\mathbf{m}_2$=[1], $\mathbf{m}_3$=[2], $\mathbf{m}_1$=[4]). Hence only node 3 can complete the service (s_3=0) destined to node 4. In this case four simultaneous job transitions take place when the job in node 3 copletes its service and moves to node 4 (i.e., at the same time the three waiting jobs move towards their destinations). The process transition is to state ((2,0,∅), (2,0,∅), (2,0,∅), (1,0,∅)) where all nodes are active, s_i=0 and $\mathbf{m}_i$=∅ $\forall i$, and n_i does not change for any node i, i=1, 2, 3, 4.

In order to define the unblocking discipline we define the following operations on the list $\mathbf{m}_i$:

Insert($\mathbf{m}_i$, j)   insert j in the list $\mathbf{m}_i$, according to the unblocking discipline,
Head($\mathbf{m}_i$)   returns the first element of $\mathbf{m}_i$ without changing $\mathbf{m}_i$,
Cancel($\mathbf{m}_i$)   cancels the first element of $\mathbf{m}_i$.

The process definition for BAS blocking type is more complex with respect to the other blocking mechanisms because process state transitions may correspond to simultaeous job transitions between the nodes, as described by examples 4.1 and 4.2. Table 4.2 shows the non-zero elements of the process transition rate matrix $\mathbf{Q}_{BAS}$ = $\|q(\mathbf{S},\mathbf{S}')\|$ for each pair of states $\mathbf{S}$, $\mathbf{S}' \in E_{BAS}$.

The indicator functions $\delta(\cdot)$ and $b_i(\cdot)$ are defined by formula (4.5) in Section 4.1.1 respectively to consider possible empty sending nodes and full destination nodes that do not allow a process transition to occur.

Table 4.2. Generator matrix Q_{BAS}

q(S,S')	relation between S and S'
$\lambda\, p_{0j}\, b_j(n_j)$	$S'_j(n'_j,s'_j,m'_j)$: $n'_j=n_j+1$, $s'_j=s_j$, $m'_j=m_j$ $S'_i=S_i\ \forall\ i\neq j$
$\delta(n_j)\,\mu_j\,f_j(n_j,K_j)\,p_{j0}$	S_j: $s_j<\min\{n_j,K_j\}$, $\mathbf{m}_j=\varnothing$ $S'_j(n'_j,s'_j,m'_j)$: $n'_j=n_j-1$, $s'_j=s_j$, $m'_j=m_j$ $S'_i=S_i\ \forall\ i\neq j$ or S_j: $s_j<\min\{n_j,K_j\}$, $\mathbf{m}_j\neq\varnothing$ S': Update(j,k) $\quad n'_k=n_k-1$
$\delta(n_j)\,\mu_j\,f_j(n_j,K_j)\,p_{ji}$	S: $s_j<\min\{n_j,K_j\}$, $\mathbf{m}_j=\varnothing$, $n_i<B_i\ i\neq j$ $S'_j(n'_j,s'_j,m'_j)$: $n'_j=n_j-1$, $s'_j=s_j$, $m'_j=m_j$ $S'_i(n'_i,s'_i,m'_i)$: $n'_i=n_i+1$, $s'_i=s_i$, $m'_i=m_i$ $S'_h=S_h\ \forall\ h\neq i,j$ or S: $s_j<\min\{n_j,K_j\}$, $n_i=B_i\ i\neq j$ $S'_j(n'_j,s'_j,m'_j)$: $n'_j=n_j$, $s'_j=s_j+1$, $m'_j=m_j$ $S'_i(n'_i,s'_i,m'_i)$: $n'_i=n_i$, $s'_i=s_i$, $m'_i=\text{Insert}(\mathbf{m}_i, j)$ $S'_h=S_h\ \forall\ h\neq i,j$
$\delta(n_j)\,\mu_j\,f_j(n_j,K_j)\,p_{ji}b_i(n_i)$	S: $s_j<\min\{n_j,K_j\}$, $\mathbf{m}_j\neq\varnothing$ S': Update(j,k) $\quad$ if $i\neq k$ then $\quad\quad\{n'_i=n_i+1,\ n'_k=n_k-1$ $\quad\quad\}$
$-\sum\limits_{S''\in E_{BAS},S''\neq S} q(S,S'')$	$S' = S$

The first row in Table 4.2 represents an external job arrival at node j. The process transition leads to the new state S'. The only difference between the two states is that $n'_j=n_j+1$ in state S'.

The second row corresponds to a job departure from the network from node j. This transition is possible only if not all the servers of node j are blocked, i.e. if $s_j<\min\{n_j, K_j\}$. In this case we distinguish the two following cases:

(i) there is no node blocked by node j, that is $\mathbf{m}_j=\varnothing$. The process transition is into state S', where $n'_j=n_j-1$ in S'_j;

(ii)   there is at least one sending node blocked by node j, that is $m_j \neq \varnothing$. Let h be
       the first node to get unblocked. When the departure occurs from node j, a
       simultaneous job transition occurs from h to j. Then if $m_h \neq \varnothing$, this latter job
       transition yields another unblocking for the first node in the list m_h, and so
       on. This is the case of a chain of blocked nodes and we have defined the
       new state components m'_k for each upstream node k in the chain through
       an algorithmic scheme. The while-loop provides an appropriate setting of
       component m'_j, by canceling the unblocked node j (Cancel(m_j)), and of
       component s'_h by unblocking the node h ($s'_h = s_h - 1$) for each pair of
       sending-destination (h,j) nodes generating a simultaneous transition. The
       assumption of deadlock-free condition guarantees that the loop ends. The
       chain of blocked nodes ends with a node k that satisfies condition (i), that is
       with $m_k = \varnothing$. A consequence of the blocked node chain is that the job
       population is changed at node k only if k is the last blocked node in the
       chain, that is $n'_k = n_k - 1$.

       The following algorithm is used to update the state of the nodes in the chain
       from node j backward to node k:

Update (j,k):

    $k \leftarrow j$;
    while ($m_k \neq \varnothing$)
    {h=Head(m_k);
    m'_k=Cancel(m_k);
    $s'_h = s_h - 1$;
    $k \leftarrow h$;
    }

The third row of Table 4.2 represents a job transition from node j towards node i
when node j does not block other nodes, that is $m_j = \varnothing$ or when the node i is full, that
is $n_i = B_i$. The job departure is possible only if not all the servers of node j are
blocked, that is if $s_j < \min\{n_j, K_j\}$. We distinguish the two following cases:

(i) $m_j = \varnothing$ and node i is not full $n_i < B_i$. In this case, the components S'_j and S'_i
       are different from those of state S and in particular $n'_j = n_j - 1$ and $n'_i = n_i + 1$.

(ii)   the destination node i is full, that is $n_i = B_i$. In this case no job transition
       occurs, but node j becomes blocked and it has to be inserted in the list of
       nodes blocked on i. Hence in the new state S' S'_j and S'_i are changed as
       follows: $s'_j = s_j + 1$ and m'_i=Insert(m_i, j).

The fourth row of Table 4.2 concerns a job transition from node j towards node
i when there are sending nodes blocked by node j, that is $m_j \neq \varnothing$. As in the previous
case, the job departure is possible only if not all node j servers are blocked, i.e.
$s_j < \min\{n_j, K_j\}$.

Let h denote the first node to be unblocked by node j according to the unblocking discipline. When a departure occurs from node j a simultaneous job transition occurs from h to j. If $m_h \neq \emptyset$, this yields another unblocking mechanism for the first node in m_h, and consequently another simultaneous job transition, and so on. The chain of blocked nodes is similar to the case of a departure from the network as discussed above for the second row. The only difference is that if the last blocked node in the chain is node i itself, node i population does not change. This is the case of a loop of blocked node, as in example 4.2 for node j=3 and node i=4.

Finally, similarly to RS-RD blocking, the last row of Table 4.2 defines the diagonal entry of the generator matrix Q_{BAS}.

4.1.4 Blocking Before Service - Server Occupied

By definition of BBS blocking, when a job begins a service (enters in a server) it declares its destination, therefore the node i state definition has to include the destination of job in service.

Let $Dest_i$ denote the set of the destinations of node i, that is:

$$Dest_i = \{j \mid p_{ij} > 0,\ 1 \leq i, j \leq M\}$$

For the sake of simplicity, let us define the following Boolean function:

$$Blocked(i) = \begin{cases} true & \text{if } \exists\, j \in Dest_i \mid B_j < N \\ false & \text{if } B_j \geq N\ \forall\, j \in Dest_i \end{cases}$$

In other words Blocked(i) is true if node i can be blocked, false otherwise. Then, node i state definition is $S_i = \begin{cases} (n_i, NS_i) & \text{if } Blocked(i) = true \\ (n_i) & \text{if } Blocked(i) = false \end{cases}$

where:
- n_i is the number of jobs in node i;
- NS_i is a vector with at most $\min\{K_i, Dest_i + 1\}$ elements and whose component $NS_{i,k}$ denotes the number of node i servers that are servicing jobs destined to node k, with $k \in Dest_i$ and for an open network $NS_{i,0}$ denotes the number of node i servers with jobs that will leave the network. By definition the following constraint holds:

$$\sum_{k \in Dest_i \cup \{0\}} NS_{i,k} = \min\{n_i, K_i\}.$$

NS_i is not defined for nodes that cannot be blocked.

For a closed network with N customers, the state space $E_{BBS\text{-}SO}$ is defined as follows:

$$E_{BBS-SO} = \{(S_1,S_2,...,S_M) \mid S_i = (n_i, NS_i) \; a_i \leq n_i \leq B_i, \; 1 \leq i \leq M, \; \sum_{i=1}^{M} n_i = N\}$$

and for an open network it is defined as follows:

$$E_{BBS-SO} = \{(S_1,S_2,...,S_M) \mid S_i = (n_i, NS_i) \; 0 \leq n_i \leq B_i, \; 1 \leq i \leq M\}.$$

Each process state transition corresponds to one of the following cases:
- an external arrival at a given node
- a departure from the network from a given node
- a transition between pair of nodes

In BBS-SO, when a job transition saturates node i all the nodes currently servicing jobs destined to i become blocked simultaneously. When a job departure from a full node i occurs, it cause the services of all upstream nodes with jobs destined to i in the servers to resume serving. To represent this system behavior, the state of the node includes the destination of the jobs currently in service.

In BBS, a job chooses its destination upon entering service.

Example 4.3. Consider three nodes of a network as depicted in Figure 4.3 where, for the sake of clarity, the label of each busy server denotes the destination of the job currently in service. We consider two servers for nodes i, j and k, i.e. $K_i=K_j=K_k=2$ and a state with $n_i>K_i$, $n_j \geq K_j$ and $n_k=0$ jobs. The two jobs in service in node i have destination nodes j and k, respectively, and the two jobs in service in node j have destination h and g, respectively.

When for example the job destined to node j ends its service in node i, the next job in queue i ($n_i>K_i$) starts the service and chooses its destination x. This corresponds to a state transition with rate $\mu_i \; f_i(n_i,K_i) \; p_{ix}$. When the job destined to node k ends its service in node i, not only the next job in queue i starts the service and chooses its destination x, but also the job moving to node k and entering the free server chooses its next destination y. Therefore, the state transition has rate $\mu_i \; f_i(n_i,K_i) \; p_{ix} \; p_{ky}$.

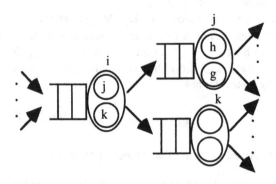

Figure 4.3. Destination choice according to BBS mechanism

Note that unlike the processes for other blocking mechanisms, the definition of the transition rates of the process for a BBS network includes routing probabilities that often concern the destination of jobs entering the servers rather than destinations of the jobs departing from the nodes.

Table 4.3 shows the non-zero entries of the infinitesimal generator $Q_{BBS-SO} = \|q(S,S')\|$ of the Markov process for the BBS-SO blocking.

For the sake of simplicity, in the table we specify only the components of S' modified by the transition, while all the other components not specified are the same as in state S. Moreover if p_{xy} appears in a transition rate definition, we implicitly assume that $y \in Dest_x \cup \{0\}$. As discussed above, for open networks the choice of the destination includes the possible destinations inside the network ($y \in Dest_x$) and a possible departure from the network ($y=0$).

We shall now discuss the definition for each entry of Table 4.3 in terms of process state transitions. Functions $\delta(\cdot)$ and $b_i(\cdot)$ are defined by (4.5) to consider possible empty sending nodes and full destination nodes that do not allow a process transition to occur. Let $z_i(n_i)$ denote the number of servers occupied by customers at node i, that is $z_i(n_i) = \min (n_i, K_i)$.

The first row in Table 4.3 corresponds to an external job arrival at node j. We further distinguish two cases depending on whether node j is blocked (Blocked(j)=true) or not. If blocked and the job enters the server, then the transition rate includes routing probability p_{jh} (where h is the next destination node) and the components n'_j and $NS'_{j,h}$ are both incremented by one, while all the other components are unchanged.

The second row corresponds to a job departure from the network from node j when the node cannot be blocked. The transition is simply the same as in a network with no blocking and the only change in the process state concerns node j state S'_j in particular $n'_j = n_j - 1$.

The third row concerns a job departure from the network from node j when node j may be blocked (Blocked(j)=true). The service rate is $NS_{j,0}\, \mu_j\, f_j(n_j,K_j)/z_j(n_j)$ where $NS_{j,0}$ is the number of jobs in service in j and destined to leave the network and $\mu_j\, f_j(n_j,K_j)/z_j(n_j)$ is the service rate of a server of node j. We further distinguish if queue j is not empty ($n_j > K_j$) and the destination of the job entering the free server in j has to be chosen (routing probability p_{jh}). After the departure, the only changes in the process state concern node j state S'_j and in particular $n'_j = n_j - 1$, $NS'_{j,0} = NS_{j,0} - 1$, $NS'_{j,h} = NS_{j,h} + 1$.

The fourth row corresponds to a job transition from node j towards node i, when node j can never be blocked (Blocked(j)=false). We further distinguish the case when the job arriving in i has not to choose its destination either because node i can never be blocked (Blocked(i)=false) or because the job arrives in the queue ($n_i \geq K_i$). In the latter case, the process transition is into state S' where only the components S'_j and S'_i are changed as $n'_j = n_j - 1$ and $n'_i = n_i + 1$. Note that since node j can never be blocked, the factor p_{ji} in the transition rate is the routing probability of the departing job that selects its destination.

Table 4.3. Generator matrix Q_{BBS-SO}

	q(S,S')	relation between S and S'
1	$\lambda\ p_{0j}\ b_j(n_j)$	S: $n_j \geq K_j \vee \neg$ Blocked(j) S': $n'_j = n_j + 1$
1	$\lambda\ p_{0j}\ b_j(n_j)p_{jh}$	S: $n_j < K_j \wedge$ Blocked(j) S': $n'_j = n_j + 1$, $NS'_{j,h} = NS_{j,h} + 1$
2	$\delta(n_j)\ \mu_j\ f_j(n_j,K_j)\ p_{j0}$	S: $\neg$ Blocked(j) S'_j: $n'_j = n_j - 1$
3	$NS_{j,0}\ \mu_j\ f_j(n_j,K_j)\ /\ z_j(n_j)$	S: Blocked(j) $\wedge\ n_j \leq K_j$ S'_j: $n'_j = n_j - 1$
3	$NS_{j,0}\ \mu_j\ f_j(n_j,K_j)\ p_{jh}\ /\ z_j(n_j)$	S: Blocked(j) $\wedge\ n_j > K_j$ S': $n'_j = n_j - 1$, $NS'_{j,0} = NS_{j,0} - 1$, $NS'_{j,h} = NS_{j,h} + 1$
4	$\delta(n_j)\ \mu_j\ f_j(n_j,K_j)\ p_{ji}$	S: $\neg$ Blocked(j) $\neg$ Blocked(i) $\vee$ (Blocked(i) $\wedge\ n_i \geq K_i$) S': $n'_j = n_j - 1$, $n'_i = n_i + 1$
4	$\delta(n_j)\ \mu_j\ f_j(n_j,K_j)\ p_{ji}\ p_{ik}$	S: $\neg$ Blocked(j) Blocked(i) $\wedge\ n_i < K_i$ S': $n'_j = n_j - 1$, $n'_i = n_i + 1$, $NS'_{i,k} = NS_{i,k} + 1$
5	$NS_{j,i}\ \mu_j\ f_j(n_j,K_j)\ b(n_i)\ /\ z_j(n_j)$	S: Blocked(j) $\wedge\ n_j \leq K_j$ $\neg$ Blocked(i) $\vee$ (Blocked(i) $\wedge\ n_i \geq K_i$) S': $n'_j = n_j - 1$, $NS'_{j,i} = NS_{j,i} - 1$ $n'_i = n_i + 1$
5	$NS_{j,i}\ \mu_j\ f_j(n_j,K_j)\ p_{jh}\ b(n_i)\ /\ z_j(n_j)$	S: Blocked(j) $\wedge\ n_j > K_j$ $\neg$ Blocked(i) $\vee$ (Blocked(i) $\wedge\ n_i \geq K_i$) S': $n'_j = n_j - 1$, $NS'_{j,i} = NS_{j,i} - 1$, $NS'_{j,h} = NS_{j,h} + 1$ $n'_i = n_i + 1$
6	$NS_{j,i}\ \mu_j\ f_j(n_j,K_j)\ p_{ik}\ /\ z_j(n_j)$	S: Blocked(j) $\wedge\ n_j \leq K_j$ Blocked(i) $\wedge\ n_i < K_i$ S': $n'_j = n_j - 1$, $NS'_{j,i} = NS_{j,i} - 1$ $n'_i = n_i + 1$, $NS'_{i,k} = NS_{i,k} + 1$
6	$NS_{j,i}\ \mu_j\ f_j(n_j,K_j)\ p_{jh}\ p_{ik}\ /\ z_j(n_j)$	S: Blocked(j) $\wedge\ n_j > K_j$ Blocked(i) $\wedge\ n_i < K_i$ S': $n'_j = n_j - 1$, $NS'_{j,i} = NS_{j,i} - 1$, $NS'_{j,h} = NS_{j,h} + 1$ $n'_i = n_i + 1$, $NS'_{i,k} = NS_{i,k} + 1$
7	$-\sum_{S'' \in E_{BBS-SO}, S'' \neq S}\ q(S,S'')$	S' = S

When node j can never be blocked then the job going from node j and arriving in i has to choose its destination because node i may be blocked (Blocked(i)=true) and the job enters a free server of i ($n_i<K_i$). Then the transition rate includes factor p_{ji}, the routing probability of the departing job that selects its destination, and p_{ik} is the routing probability of the job arriving in i that chooses its next destination. Hence in the new process state **S** there are not only new components n'_j and n'_i like the previous case, but also an increment of the number of jobs in service in node i destined to k, i.e. $NS'_{i,k}=NS_{i,k}+1$.

The following two rows (fifth and sixth) in Table 4.3, represent a job transition from node j to node i when node j may be blocked (Blocked(j)=true). The service rate is $NS_{j,i}\ \mu_j f_j(n_j,K_j)/z_j(n_j)$ where $NS_{j,i}$ is the number of jobs in service in j and destined to i.

The fifth row corresponds to the case that the job arriving in i has not to select the destination either because node i can never be blocked (Blocked(i)=false), or because the job arrives in the queue ($n_i\geq K_i$). Since node j may be blocked, we further distinguish two cases depending on whether the queue of node j is empty ($n_j\leq K_j$). If the queue j is not empty ($n_j>K_j$) a job enters a free server in j and chooses its next destination (with routing probability p_{jh}). The new state **S'** has modified components **S'$_j$** and **S'$_i$**: $n'_j=n_j-1$, $NS'_{j,h}=NS_{j,h}+1$ and $n'_i=n_i+1$. If the queue j is empty ($n_j\leq K_j$) the transition rate does not include routing probability p_{jh} and component $NS'_{j,h}$ is not modified ($NS'_{j,h}=NS_{j,h}$).

Then the sixth row of Table 4.3 corresponds to the case the job arriving in i chooses its destination because node i may be blocked (Blocked(i)=true) and the job enters a free server in i ($n_i<K_i$). The transition rate includes p_{ik} that is the routing probability of the destination of the job arriving in i. We consider the two cases: queue j is empty ($n_j\leq K_j$) or not ($n_j>K_j$). If queue j is not empty, a job enters a free server in j and chooses its next destination (with routing probability p_{jh}) and the next state **S'** has the new components **S'$_j$** and **S'$_i$** with $n'_j=n_j-1$, $NS'_{j,h}=NS_{j,h}+1$ and $n'_i=n_i+1$, $NS'_{i,k}=NS_{i,k}+1$. If the queue j is empty ($n_j\leq K_j$), the transition rate does not include routing probability p_{jh} and component $NS'_{j,h}$ is not modified. Note that both the routing probabilities p_{ik} and p_{jh} are referred to the next destinations of jobs entering the service.

Finally, the last row of Table 4.3 defines the diagonal entry of the generator matrix $\mathbf{Q}_{BBS-SO}$.

4.1.5 Blocking Before Service - Server Not Occupied

As discussed in Chapter 2 the mechanism BBS-SNO can only be defined for networks with special topologies, i.e. if node i has finite capacity then there exists only one node j such that $p_{ji}>0$ and $p_{ki}=0$ for $k\neq j$, $1\leq i,j,k\leq M$.

The process state definition for this blocking mechanism is identical to the definition given for BBS-SO, but the state space $E_{BBS-SNO}$ is different from E_{BBS-SO} because the states with $n_i=B_i$, $n_h=B_h$ with $h\in Dest_i$ do not belong to the $E_{BBS-SNO}$ when node i is blocked. In particular when node i is blocked the jobs blocked in it stay in the queue and a transition towards node i may be defined only if there is effective room in queue i. This means that when node i may be blocked, that is Blocked(i)=true, the job transition from node j to node i occurs only if the following further condition holds:

$$n_i < B_i - \sum_{z \in Dest_i, n_z = B_z} NS_{i,z} \qquad (4.6)$$

where the right-hand term represents the free waiting room in queue i. The definition of the process transition rate matrix $Q_{BBS-SNO}$ is identical to that of matrix Q_{BBS-SO} given in the previous section in Table 4.3, except condition (4.6) has to be satisfied in the fourth and fifth rows.

4.1.6   Blocking Before Service - Overall

According to the Overall Blocking Before Service mechanism, when a destination node i becomes full, it blocks the service in each of its possible sending nodes j, regardless of the destination of the currently processed job. For this reason, node i state definition does not include the destinations of jobs in service as in other BBS cases. Therefore node i state definition is simply $S_i = n_i$. Note that the state definition is identical to the state definition of the process for the network with RS-RD mechanism. However, the state meaning and the state space definition of the two processes are different, as described next.

For all blocking mechanisms RS-RD, RS-FD, BAS, BBS-SO and BBS-SNO a job is blocked, according to the different rules, only if it is directed to a full node. In the case of BBS-O, a job in node j is blocked as soon as any destination node i ($i \in Dest_j$) becomes full, regardless of the actual destination of the job currently in service (that is, even if this job is destined to another node g ($g \in Dest_j$) which is not full). The blocking mechanism in BBS-O is the most restrictive one because server activity is blocked more frequently than with other mechanisms.

Let us define the following function for any state $S=(n_1, n_2,..., n_M)$:

$$bl_j(S)= \begin{cases} i & \text{if node j is blocked by node i} \\ 0 & \text{if node j is active} \end{cases}$$

Then a state S with a full node i ($n_i=B_i$) is feasible only if there is an upstream node j of node i that is blocked by node i, that is

$$n_i=B_i \Rightarrow \exists j : i \in Dest_j, bl_j(S) = i \qquad (4.7)$$

In other words a state S is feasible if it is possible that this condition is satisfied.

Example 4.4. Consider for example, the portion of network shown in Figure 4.4. Consider state $S=(...,B_i,...,B_g,...,B_h,...)$ where destination nodes i, g and h are full. State S is not feasible because for any value of functions $bl_j(S)$ and $bl_k(S)$ of the sending nodes, nodes i, g and h cannot be simultaneously full, i.e. the three nodes cannot simultaneously satisfy condition (4.7). Indeed, for example if $bl_j(S)=i$ and $bl_k(S)=h$, node g cannot become full because it does not have active upstream node. For any possible value of the functions there is at least one node that cannot be full.

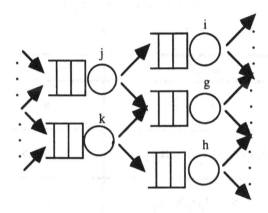

Figure 4.4. Example 4.4

The process state space E_{BBS-O} for a closed network with N customers is defined as follows:

$$E_{BBS-O} = \{(n_1, n_2,..., n_M) \mid a_i \leq n_i \leq B_i, \sum_{i=1}^{M} n_i = N \text{ and (4.7) holds for } 1 \leq i \leq M\}$$

and for an open networks it is defined as follows:

$$E_{BBS-O} = \{(n_1, n_2,..., n_M) \mid 0 \leq n_i \leq B_i \text{ and (4.7) holds for } 1 \leq i \leq M\}.$$

For this blocking mechanism, as discussed above, the blocking of the servers occurs more frequently than for the other mechanisms. As soon as a job transition saturates a node i, all its sending nodes become blocked regardless of the destination of the job currently in service. Some job transitions correspond to the unblocking phenomenon of several nodes: for example if a set of nodes has only one full destination node i, then as soon as a job leaves node i all the sending nodes become unblocked. To define the process transition rates we define the following function $\beta_j(S)$ to indicate whether node j is blocked when the network is in state S, for $S \in E_{BBS-O}$, $1 \leq j \leq M$:

$$\beta_j(S) = \begin{cases} 0 & \text{if } bl_j(S) \neq 0 \\ 1 & \text{if } bl_j(S) = 0 \end{cases} \qquad (4.8)$$

Table 4.4 shows the non-zero transition rates of the process infinitesimal generator $Q_{BBS-O} = \|q(S,S')\|$ for each pair S, $S' \in E_{BBS-O}$.

In Table 4.4 the indicator functions $\delta(\cdot)$ and $b_i(\cdot)$ are defined by (4.5) in Section 4.1.1 respectively to refer to empty sending nodes and full destination nodes that do not allow a process transition to occur. Expression μ_j $f_j(n_j, K_j)$ denotes the job service rate of node j with K_j servers and n_j jobs in it. Finally, e_i denotes the M-vector with all zero components except one in i-th position.

Table 4.4. Generator matrix Q_{BBS-O}

q(S,S')	relation between S and S'
$\lambda\ p_{0j}\ b_j(n_j)$	$S' = S + e_j$
$\delta(n_j)\ \beta_j(S)\ \mu_j\ f_j(n_j, K_j)\ p_{j0}$	$S' = S - e_j$
$\delta(n_j)\ \beta_j(S)\ \mu_j\ f_j(n_j, K_j)\ p_{ji}\ b_i(n_i)$	$S' = S + e_i - e_j,\ i \neq j$
$-\ \sum\limits_{S'' \in E_{BBS-O}, S'' \neq S}\ q(S,S'')$	$S' = S$

The first row of Table 4.4 represents an external job arrival at node j that enters the node only if $n_j < B_j$ and this is considered by function $b_j(n_j)$. The new process state S' has a changed component S'_j for the node j state and in particular $n'_j = n_j + 1$.

The second row concerns a job departure from the network from node j that occurs only if $n_j > 0$ (corresponding to function $\delta(n_j)$) and if node j is not blocked (corresponding to function $\beta_j(S)$). The only change in the new process state concerns the node j state S'_j with $n'_j = n_j - 1$.

The third row represents a job transition from node j towards node i. This transition is possible only if node j is not empty and not blocked and node i is not full and this correspond to functions $\delta(n_j)$, $\beta_j(S)$ and $b_i(n_i)$, respectively. The state change concerns only node j and node i populations, in particular $n'_j = n_j - 1$ and $n'_i = n_i + 1$.

Finally, the last row of Table 4.4 defines the diagonal entry of the generator matrix Q_{BBS-O}.

4.1.7 A simple example

To illustrate the model definition for the various cases described in the previous sections, we shall now present a simple example of a closed network and the underlying Markov processes for different blocking mechanisms. Consider the three node closed network with central server topology shown in Figure 4.5. We show the state transition diagram of the Markovian process underlying the network model. We compare the state transition diagrams of the network models with RS-RD, BBS-SO, BAS and BBS-O mechanisms.

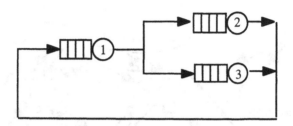

Figure 4.5. A closed network

We consider a network with finite capacities $B_1=B_2=B_3=2$, single server nodes ($K_i =1 \; \forall i$) and N=3 jobs. For the sake of clarity, whenever possible we simplify the complete state notation defined in previous Sections.

The state of the network with RS-RD mechanism is simple $S = (n_1,n_2,n_3)$ and

$$E_{RS-RD} = \{(2, 1, 0), (2, 0, 1), (1, 2, 0), (1, 1, 1), (1, 0, 2), (0, 2, 1), (0, 1, 2)\}$$

Figure 4.6 shows the state transition diagram where an edge from state **S** to state **S'** represents the transition between the two connected states. For the sake of readability, we show the labels of the edges apart from the diagram in Figure 4.6. They correspond to four transition types defined in the third row of Table 4.1.

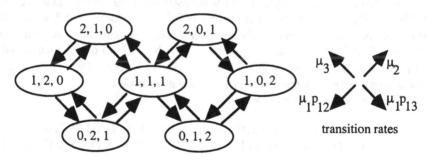

Figure 4.6. State transition diagram of the network with RS-RD blocking

The network with BAS blocking has a state $S = (S_1,S_2,S_3)$ where $S_i=(n_i,s_i,m_i)$ $\forall i$ and the following state space:

$$E_{BAS} = \{(2, 1, 0), ((2,[2]), (1,1), 0), (2, 0, 1), ((2,[3]), 0, (1,1)), (1, 2, 0), ((1,1), (2,[1]), 0), (1, 1, 1), (1, 0, 2), ((1,1), 0, (2,[1])), (0, 2, 1), (0, 1, 2)\}$$

We simplify the state notation by including only the essential components, while the reader can easily derive the others. The state transition diagram is shown in Figure 4.7 together with the four types of transition rates, defined in the third row of Table 4.2.

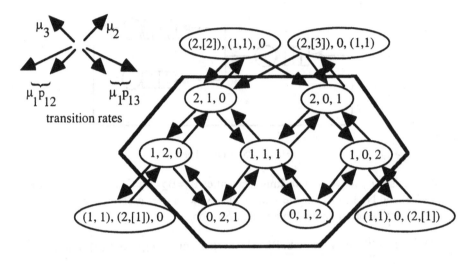

Figure 4.7. State transition diagram of the network with BAS blocking

Note that if we discard the four states with a blocked node and the transitions to and from those states, as illustrated in Fig. 4.7, then the transition diagram becomes identical to the diagram shown in Figure 4.6 for the network with RS-RD blocking.

For the network working under BBS-SO mechanism the state of node i is given by $S_i=(n_i,NS_i)$ $\forall i$. Note that when node 1 is full, then nodes 2 and 3 are blocked since they have node 1 as their only destination. Hence we omit the state component NS_i for i=2, 3. Moreover since $K_1=1$ the set NS_1 has at most one element equal to $1_{1,k}$, that means the server of node 1 has in service a job destined to node k (k=2, 3). For simplicity, we use k instead of $1_{1,k}$ in the state of node 1.

The process state space is the following:

$$E_{BBS-SO} = \{((2,2), 1\ 0), ((2,3), 1, 0), (2,2), 0, 1), ((2,3), 0, 1), ((1,2), 2, 0), ((1,3), 2, 0), ((1,2), 1, 1), ((1,3), 1, 1), ((1,2), 0, 2), ((1,3), 0, 2), (0, 2, 1), (0, 1, 2)\}$$

It is clear that this case cannot be easily compared with the others since each state with $n_1>0$ is duplicated into the two states with k=2 and k=3. In Figure 4.8 we show only a portion of the state transition diagram for state ((1,2),1,1) and the related transitions. The six types of transition rates are defined by the third through sixth rows of Table 4.3.

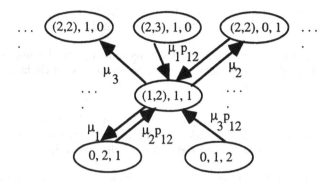

Figure 4.8. Portion of the state transitions of the network with BBS-SO blocking

Finally, for the network with BBS-O mechanism the state is simply given by $S=(n_1,n_2,n_3)$ and the state space is identical to that of the network with RS-RD blocking:

$$E_{BBS-O} = \{(2, 1, 0), (2, 0, 1), (1, 2, 0), (1, 1, 1), (1, 0, 2), (0, 2, 1), (0, 1, 2)\}$$

However, the state transition diagram shown in Figure 4.9 differs from the network with RS-RD blocking, because for the BBS-O the overall mechanism defines that there are no transitions from node 1 when one of its destinations is full, i.e. from states (1,2,0) and (1,0,2). The transition rates are defined by the third row of Table 4.4.

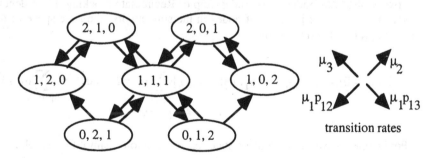

Figure 4.9. State transition diagram of the network with BBS-O blocking

4.1.8 Stop and Recirculate Blocking

For these two blocking types the population of a subnetwork or the network is assumed to be in the range [L,U], where L and U are the minimum and maximum population admitted, respectively. This constraint can be represented by an appropriate definition of both the load dependent arrival rate function a(n), n≥0 and of the (network) blocking function d(n), when n jobs are present in the network n≥0. Indeed, the following constraints hold:

$a(U)=0$ if $U<\infty$ and $a(n)>0$ for $L\leq n<U$,
$d(L)=0$ if $L>0$ and $0<d(n)\leq 1$ for $L<n\leq U$.

For multichain networks arrival and blocking functions can be also defined for each chain r, dependent on the total network population N_r in chain r, $a_r(N)$ and $d_r(N)$, $N=(N_1,\ldots, N_R)\geq 0$, $1\leq r\leq R$.

For Stop and Recirculate blocking the definition of node i state is simply $S_i=n_i$, similar to RS-RD, BBS-O and networks with infinite capacity queues. However, the state meaning and the state space definition of the two relative processes are different.

In a network with Stop blocking the service rate of each node is delayed by a factor $0<d(n)\leq 1$, where $L<n\leq U$ is the total network (or subnetwork) population. All the servers are blocked when the total network population $n=n_1+\ldots+n_M$ reaches its minimum value (n=L) for which the (network) blocking function yields d(n)=0. The service in the nodes is resumed as soon as an arrival modifies the total population.

In Recirculate blocking the servers are always active. A job upon completion of its service at queue j actually leaves the network with probability $p_{j0}d(n)$, when n is the total network population, while with probability $p_{j0}[1-d(n)]$ it stays in the network, and moves to a destination node according to the routing probabilities. Consequently, a job completing the service at node j actually enters node i with state dependent routing probability $(p_{ji}+p_{j0}[1-d(n)]p_{0i})$, with $1\leq i,j\leq M$, $n\geq 0$. When the total network population reaches its minimum value (n=L) the routing probabilities are modified to force the jobs to recirculate into the network (d(n)=0).

Let us consider a closed network with M nodes and N customers, in which a subnetwork of $M'\leq M$ nodes work under Stop or Recirculate blocking. If we denote by $Sub=\{i_1, i_2, \ldots, i_{M'}\}$ the set of nodes in the subnetwork, the state space of the system E_{Stop} is defined as follows:

$$E_{Stop} = E_{Recirculate} = \{(n_1, n_2,\ldots, n_M) \mid \sum_{i=1}^{M} n_i = N \text{ and } L \leq \sum_{i\in Sub} n_i \leq U, \ n_i\geq 0, \text{ for } 1\leq i\leq M\}$$

For an open network, the process state space definition is simplified as follows:

$$E_{Stop} = E_{Recirculate} = \{(n_1, n_2,\ldots, n_M) \mid L \leq \sum_{i\in Sub} n_i \leq U, n_i\geq 0, \text{ for } 1\leq i\leq M\}$$

The process state transitions correspond to the following cases:

- an external arrival at a given node that enters the network
- a departure from the network from a given node
- a transition between a given pair of nodes.

The last type of state transition in a network with Recirculate blocking can correspond to the blocking phenomenon, as follows: when a job attempts to exit the network when there is the minimum population in it, the job is forced to recirculate within the network.

Tables 4.5 and 4.6 show the non-zero components of the transition rate matrices $Q_{Stop} = \|q(S,S')\|$, for $S,S' \in E_{Stop}$ and $Q_{Recirculate} = \|q(S,S')\|$, for $S,S' \in E_{Recirculate}$. We assume that the arrival rate function $a(n)$ and the blocking function $d(n)$ are equal to 1 ($\forall$ n), if the arrival node and the departure node do not belong to the subnetwork *Sub*. The first three rows of both tables correspond to the three transition types introduced above and the last row defines the diagonal element.

Table 4.5. Generator matrix Q_{Stop}

$q(S,S')$	relation between S and S'
$\lambda\, p_{0j}\, a(n)$	$S' = S + e_j$ if $a(n) \neq 0$ $S' = S$ otherwise
$\delta(n_j)\, \mu_j\, f_j(n_j, K_j)\, d(n)\, p_{j0}$	$S' = S - e_j$ if $d(n) \neq 0$ $S' = S$ otherwise
$\delta(n_j)\, \mu_j\, f_j(n_j, K_j)\, d(n)\, p_{ji}$	$S' = S + e_i - e_j$ if $d(n) \neq 0$, $i \neq j$ $S' = S$ otherwise
$- \displaystyle\sum_{S'' \in E_{Stop},\, S'' \neq S} q(S,S'')$	$S' = S$

Table 4.6. Generator matrix $Q_{Recirculate}$

$q(S,S')$	relation between S and S'
$\lambda\, p_{0j}\, a(n)$	$S' = S + e_j$ if $a(n) \neq 0$ $S' = S$ otherwise
$\delta(n_j)\, \mu_j\, f_j(n_j, K_j)\, d(n)\, p_{j0}$	$S' = S - e_j$ if $d(n) \neq 0$ $S' = S$ otherwise
$\delta(n_j)\, \mu_j\, f_j(n_j, K_j)\, (p_{ji} + p_{j0}\,[1 - d(n)]\, p_{0i})$	$S' = S + e_i - e_j$ $i \neq j$
$- \displaystyle\sum_{S'' \in E_{Recirculate},\, S'' \neq S} q(S,S'')$	$S' = S$

The indicator function $\delta(\cdot)$ is defined by (4.5) in Section 4.1.1 to consider possible empty sending nodes that do not allow a process transition to occur and μ_j $f_j(n_j, K_j)$ denotes the job service rate of node j with K_j servers and n_j jobs in it.

The first row of the two tables concerns an external job arrival at node j that enters the node only if the networks has not reached its maximum capacity U which is represented by factor a(n) in the transition rate. This transition results in the new state **S'** with n'$_j$=n$_j$+1.

The second row represents a job departure from the network from node j that occurs only if n$_j$>0 (i.e. δ(n$_j$)>0) and if the network has not reached its minimum capacity L (i.e. d(n)>0) for both Stop and Recirculate blocking. This transition results in the new state with n'$_j$=n$_j$−1.

The third row corresponds to a job transition from node j towards node i, if node j is not empty (δ(n$_j$)>0). For Stop blocking, this transition is possible only if the network population is not equal to the minimum threshold L (d(n)>0), otherwise (d(n)=0) all nodes are blocked. In Recirculate blocking any internal job transition from node j to node i may occur according to probability (p$_{ji}$ + p$_{j0}$[1−d(n)]p$_{0i}$). That is, when d(n)<1, a system departure from node j "triggers" an instantaneous new arrival to node i. The state change concerns only node j and node i populations as follows: n'$_j$=n$_j$−1 and n'$_i$=n$_i$+1.

Finally, the last rows of Tables 4.5 and 4.6 define the diagonal entry of the generator matrices **Q** $_{Stop}$ and **Q** $_{Recirculate}$.

4.1.9 Heterogeneous networks

In the previous sections we have considered networks in which each node works under the same blocking types. In this case the network is called homogeneous. Complex real systems such as computer and communication systems often have different behaviors of the blocking mechanism for different components.

To represent complex systems we consider heterogeneous networks where different nodes can work under different blocking mechanisms. Note that a node j working under a particular blocking type X means that when its destination node i is full, node j behaves according to X blocking model. Of course, in this case, the process definition is more complex because more constraints of the different blocking types must be verified for a given node as upstream nodes working under different blocking types. For example, consider node i with two upstream nodes, in particular node j with BAS blocking and node k with BBS-O blocking. If node i has finite capacity, the node state notation S$_i$ must include component **m**$_i$ as defined in Section 4.1.3. On the other hand, the feasibility criterion (4.7) defined in Section 4.1.6 must be checked in states with n$_i$=B$_i$.

In this section for the sake of simplicity, we do not consider BBS-SNO, Stop and Recirculate mechanisms. One can include these blocking types in a heterogeneous network by taking into account the fraework constructed in respective sections. Recalling that for exponential queueing networks with LIFO discipline RS-FD blocking is equivalent to BBS-SO blocking, in the following we refer only to BBS-SO mechanism.

Let us define a partition of M nodes in the network as follows:

$$\bigcup_{X \in B} I_X = \{1, 2, ..., M\}$$

where $B = \{$Unb, RS-RD, BAS, BBS-SO, BBS-O$\}$, I_X is the set of the nodes with X blocking type and X=Unb denotes that a node is never blocked.

The state space of a closed heterogeneous network with M nodes operating according to a blocking type in B is defined as follows:

$$E_{Het} = \{(S_1, S_2,..., S_M) \mid$$
$$S_i = (n_i, m_i) \; \forall \; i \in I_{RS\text{-}RD} \cup I_{BBS\text{-}O} \cup I_{Unb};$$
$$S_i = (n_i, s_i, m_i) \; \forall \; i \in I_{BAS};$$
$$S_i = (n_i, NS_i, m_i) \; \forall \; i \in I_{BBS\text{-}SO};$$
$$\text{conditions (4.2) and (4.3) hold for } 1 \leq i \leq M\}$$

where components s_i, m_i and NS_i are defined as in Sections 4.1.3 and 4.1.4 and we assume $m_i = \varnothing$ if node i has not sending node with BAS blocking.

For an open network, the process state space definition is simplified by substituting conditions (4.2) and (4.3) with condition (4.1) in the state space definition.

In the above construction each job transition from node j to node i can yield simultaneous transitions if $m_j \neq \varnothing$, and it has at least an upstream node with BAS blocking and a job waiting for room in node j.

The generator matrix $Q_{Het} = \|q(S,S')\|$ is defined in Table 4.7 and it is divided according to the blocking type of the sending node j in five cases. As in previous cases of homogeneous networks, Table 4.7 defines the matrix element $q(S,S')$ for each pair $S, S' \in E_{Het}$ with $S \neq S'$, and the relation between S and S'.

For the sake of simplicity, in the Table we specify only the components of S' that change with the transition and other components not specified remain unchanged. The functions $\delta(\cdot)$ and $b(\cdot)$ are both defined as before.

In Table 4.7.a, we consider the case where node j can never be blocked or works under RS-RD blocking. Node j may have limited capacity and hence state component m_j could not be empty.

The first row concerns an external job arrival at node j. This transition happens if node j is not full $(b_j(n_j)=1)$. The only change in the process state concerns node j state where $n'_j = n_j + 1$.

The second row corresponds to a job departure from the network from node j. Each job transition from a node j could yield simultaneous transitions if node j is full and has a sending node with BAS blocking and a job waiting for room in j.
For this reason, we distinguish two cases:
(i) node j has not BAS upstream nodes blocked on it, that is $m_j = \varnothing$. As a consequence, the process transition is into state S' with $n'_j = n_j - 1$;
(ii) node j has BAS upstream nodes blocked on it, that is $m_j \neq \varnothing$. According to the unblocking discipline, a node h has to be unblocked and a simultaneous job transition from h occurs. As for the case of a homogeneous network with BAS blocking, this transition from node h may yield a series of unblocking events if a chain of blocked nodes exist. The while-loop provides the appropriate setting of component m'. A consequence of the blocked node chain is that the job

Table 4.7.a. Generator matrix Q_{Het}: $j \in I_{Unb} \cup I_{RS-RD}$

q(S,S')	relation between S and S'
$\lambda\ p_{0j}\ b_j(n_j)$	S': $n'_j = n_j + 1$
$\delta(n_j)\ \mu_j\ f_j(n_j, K_j)\ p_{j0}$	S: $m_j = \varnothing$ S': $n'_j = n_j - 1$ or S: $m_j \neq \varnothing$ S': Update (j,z) $n'_z = n_z - 1$

$\delta(n_j)\ \mu_j\ f_j(n_j, K_j)\ p_{ji}\ b_i(n_i)$	S: $m_j = \varnothing$, ($i \in I_{BBS-SO} \wedge n_i \geq K_i$)
	     $\vee\ i \notin I_{BBS-SO}$
	S': $n'_j = n_j - 1$, $n'_i = n_i + 1$
	or
	S: $m_j \neq \varnothing$, ($i \in I_{BBS-SO} \wedge n_i \geq K_i$)
	     $\vee\ i \notin I_{BBS-SO}$
	S': Update (j,z)
	   if $i \neq z$ then
	     $\{\ n'_i = n_i + 1,\ n'_z = n_z - 1$
	     $\}$

$\delta(n_j)\ \mu_j\ f_j(n_j, K_j)\ p_{ji}\ P_{ik}$	S: $m_j = \varnothing$, ($i \in I_{BBS-SO} \wedge n_i < K_i$)
	S': $n'_j = n_j - 1$, $n'_i = n_i + 1$, $NS'_{i,k} = NS_{i,k} + 1$
	or
	S: $m_j \neq \varnothing$, ($i \in I_{BBS-SO} \wedge n_i < K_i$)
	S': Update (j,z)
	   if $i \neq z$ then
	     $\{\ n'_i = n_i + 1,\ n'_z = n_z - 1$
	     $\}$
	$NS'_{i,k} = NS_{i,k} + 1$

population is changed at node z if z is the last blocked node in the chain, that is $n'_z = n_z - 1$. For further details see Section 4.1.3.

The third row corresponds to a job transition from node j to node i. Note that if $j \in I_{Unb}$ node i does not have limited capacity and $b_i(n_i) = 1$. Otherwise, if $j \in I_{RS-RD}$, node i may have limited capacity and the transition takes place only if node i is not full. In both cases, the destination node i could be blocked and therefore we have to consider the different blocking mechanism for node i.

We further distinguish two cases corresponding to node i state. In the first case node i does not work under BBS-SO or $n_i \geq K_i$. As a consequence, the destination of the job arriving in i is not chosen. The second case is when node i works under BBS-SO and the job enters a free server of i ($n_i < K_i$). In this case, the transition rate includes factor p_{ji} concerning the destination of the departing job, and factor p_{ik} concerning next destination of the job arriving in i. In both cases, the possible simultaneous transitions must be considered.

In Table 4.7.b, we consider the case where node j works under BAS blocking.
As the reader can easily see, this table shows two differences with respect to Table 4.7.a. First the condition $s_j < \min\{n_j, K_j\}$ is added in the second column. Indeed, all the transitions from node j may happen only if not all node j servers are blocked.

Second, we add the last row that represents the transition corresponding to the blocking due to BAS mechanism. This happens when the destination node i is full, that is $n_i = B_i$. In this case no job transition takes place, but the blocking of node j occurs and node j has to be inserted in the list of nodes blocked on i.

In Table 4.7.c, we consider the case where node j works under BBS-SO blocking. This table can be easily obtained from Table 4.3 in Section 4.1.4 that defines the generator for a homogeneous network with BBS-SO blocking. Table 4.7.c has less rows than the previous one because the condition Blocked(j) is always true given the assumption $j \in I_{BBS\text{-}SO}$. However, we have always to consider the simultaneous job transitions when $\mathbf{m}_j \neq \varnothing$ as in the previous cases.

In Table 4.7.d, we consider the case where node j works under BBS-O blocking.
The first row concerns an external job arrival at node j and, as in the previous cases, this transition may happen if node j is not full ($b_j(n_j)=1$).
The remaining rows of Table 4.7.d represent job transitions from node j that are possible only if node j is not blocked. According to the BBS-O blocking mechanism, this is considered by function $\beta_j(\mathbf{S})$ defined by expression (4.8) in Section 4.1.6.
As for the previous cases, we have to consider the simultaneous transitions if node j is full and has an upstream node with BAS blocking and a job waiting for room in j.
Moreover, we have to consider the different blocking mechanism for node i and in particular, if node i works under BBS-SO and the job enters a free server of i ($n_i < K_i$), we have to determine the next destination of the job arriving in i (factor p_{ik}).

Finally, Table 4.7.e shows the usual definition of the diagonal entry of the generator matrix $\mathbf{Q}_{Het}$.

4.2 QUEUE LENGTH DISTRIBUTION AND AVERAGE PERFORMANCE INDICES

The presentation order of the blocking types in the previous sections corresponds to the increasing order of the blocking phenomenon occurrence. Indeed, in a network with RS mechanism we observe the least blocking, because, by definition all servers are always active and a customer that cannot enter a saturated node receives a new

Table 4.7.b. Generator matrix Q_{Het}: $j \in I_{BAS}$

q(S,S')	relation between S and S'
$\lambda\ p_{0j}\ b_j(n_j)$	**S':** $n'_j=n_j+1$
$\delta(n_j)\ \mu_j\ f_j(n_j,K_j)\ p_{j0}$	**S:** $s_j<\min\{n_j, K_j\}$, $\mathbf{m_j}=\varnothing$ **S':** $n'_j=n_j-1$ or **S:** $s_j<\min\{n_j, K_j\}$, $\mathbf{m_j}\neq\varnothing$ **S':** Update (j,z) $n'_z=n_z-1$
$\delta(n_j)\ \mu_j\ f_j(n_j,K_j)\ p_{ji}\ b_i(n_i)$	**S:** $s_j<\min\{n_j, K_j\}$, $\mathbf{m_j}=\varnothing$, $(i\in I_{BBS\text{-}SO} \wedge n_i\geq K_i)$ $\vee\ i\notin I_{BBS\text{-}SO}$ **S':** $n'_j=n_j-1$, $n'_i=n_i+1$ or **S:** $s_j<\min\{n_j, K_j\}$, $\mathbf{m_j}\neq\varnothing$, $(i\in I_{BBS\text{-}SO} \wedge n_i\geq K_i)$ $\vee\ i\notin I_{BBS\text{-}SO}$ **S':** Update (j,z) if $i\neq z$ then $\{\ n'_i=n_i+1,\ n'_z=n_z-1$ $\}$
$\delta(n_j)\ \mu_j\ f_j(n_j,K_j)\ p_{ji}\ p_{ik}$	**S:** $s_j<\min\{n_j, K_j\}$, $\mathbf{m_j}=\varnothing$, $(i\in I_{BBS\text{-}SO} \wedge n_i<K_i)$ **S':** $n'_j=n_j-1$, $n'_i=n_i+1$, $NS'_{i,k}=NS_{i,k}+1$ or **S:** $s_j<\min\{n_j, K_j\}$, $\mathbf{m_j}\neq\varnothing$, $(i\in I_{BBS\text{-}SO} \wedge n_i<K_i)$ **S':** Update (j,z) if $i\neq z$ then $\{\ n'_i=n_i+1,\ n'_z=n_z-1$ $\}$ $NS'_{i,k}=NS_{i,k}+1$
$\delta(n_j)\ \mu_j\ f_j(n_j,K_j)\ p_{ji}$	**S:** $s_j<\min\{n_j, K_j\}$, $n_i=B_i$ **S':** $s'_j=s_j+1$, $\mathbf{m'_i}=\text{Insert}(\mathbf{m_i}, j)$

Table 4.7.c. Generator matrix Q_{Het}: $j \in I_{BBS-SO}$

q(**S**,**S'**)	relation between **S** and **S'**
$\lambda \, p_{0j} \, b_j(n_j)$	**S**: $n_j \geq K_j$ **S'**: $n'_j = n_j + 1$
$\lambda \, p_{0j} \, b_j(n_j) \, p_{jh}$	**S**: $n_j < K_j$ **S'**: $n'_j = n_j + 1$, $NS'_{j,h} = NS_{j,h} + 1$
$NS_{j,0} \, \mu_j \, f_j(n_j, K_j)$	**S**: $n_j \leq K_j \wedge \mathbf{m_j} = \varnothing$ **S'**: $n'_j = n_j - 1$

$NS_{j,0} \, \mu_j \, f_j(n_j, K_j) \, p_{jh}$	**S**: $n_j > K_j \wedge \mathbf{m_j} = \varnothing$
	S': $n'_j = n_j - 1$, $NS'_{j,h} = NS_{j,h} + 1$
	         or
	
	S: $\mathbf{m_j} \neq \varnothing$
	S': Update (j,z)
	     $n'_z = n_z - 1$, $NS'_{j,h} = NS_{j,h} + 1$

$NS_{j,i} \, \mu_j \, f_j(n_j, K_j) \, b_i(n_i)$	**S**: $n_j \leq K_j \wedge \mathbf{m_j} = \varnothing$, $(i \in I_{BBS-SO} \wedge n_i \geq K_i) \vee i \notin I_{BBS-SO}$ **S'**: $n'_j = n_j - 1$, $n'_i = n_i + 1$

$NS_{j,i} \, \mu_j \, f_j(n_j, K_j) \, p_{jh} \, b_i(n_i)$	**S**: $n_j > K_j \wedge \mathbf{m_j} = \varnothing$,
	    $(i \in I_{BBS-SO} \wedge n_i \geq K_i) \vee i \notin I_{BBS-SO}$
	S': $n'_j = n_j - 1$, $NS'_{j,h} = NS_{j,h} + 1$, $n'_i = n_i + 1$
	         or
	
	S: $\mathbf{m_j} \neq \varnothing$,
	    $(i \in I_{BBS-SO} \wedge n_i \geq K_i) \vee i \notin I_{BBS-SO}$
	S': Update (j,z)
	    if $i \neq z$ then { $n'_i = n_i + 1$, $n'_z = n_z - 1$
	            }
	    $NS'_{j,h} = NS_{j,h} + 1$

$NS_{j,i} \, \mu_j \, f_j(n_j, K_j) \, p_{ik}$	**S**: $n_j \leq K_j \wedge \mathbf{m_j} = \varnothing$ $(i \in I_{BBS-SO} \wedge n_i < K_i)$ **S'**: $n'_j = n_j - 1$, $n'_i = n_i + 1$, $NS'_{i,k} = NS_{i,k} + 1$

$NS_{j,i} \, \mu_j \, f_j(n_j, K_j) \, p_{jh} \, p_{ik}$	**S**: $\mathbf{m_j} \neq \varnothing$
	    $(i \in I_{BBS-SO} \wedge n_i < K_i)$
	S': Update (j,z)
	    if $i \neq z$ then { $n'_i = n_i + 1$, $n'_z = n_z - 1$
	            }
	    $NS'_{i,k} = NS_{i,k} + 1$

Table 4.7.d. Generator matrix Q_{Het}: $j \in I_{BBS-O}$

q(S,S')	relation between S and S'
$\lambda\ p_{0j}\ b_j(n_j)$	S': $n'_j=n_j+1$
$\delta(n_j)\ \beta_j(S)\ \mu_j\ f_j(n_j,K_j)\ p_{j0}$	S: $m_j=\emptyset$ S': $n'_j=n_j-1$ or S: $m_j\neq\emptyset$ S': Update (j,z) $n'_z=n_z-1$
$\delta(n_j)\ \beta_j(S)\ \mu_j\ f_j(n_j,K_j)\ p_{ji}\ b_i(n_i)$	S: $m_j=\emptyset$, $(i \in I_{BBS-SO} \wedge n_i \geq K_i)$ $\vee\ i \notin I_{BBS-SO}$ S': $n'_j=n_j-1,\ n'_i=n_i+1$ or S: $m_j\neq\emptyset$, $(i \in I_{BBS-SO} \wedge n_i \geq K_i)$ $\vee\ i \notin I_{BBS-SO}$ S': Update (j,z) if $i \neq z$ then $\{\ n'_i=n_i+1,\ n'_z=n_z-1$ $\}$
$\delta(n_j)\ \beta_j(S)\ \mu_j\ f_j(n_j,K_j)\ p_{ji}\ p_{ik}$	S: $m_j=\emptyset$, $(i \in I_{BBS-SO} \wedge n_i<K_i)$ S': $n'_j=n_j-1,\ n'_i=n_i+1,\ NS'_{i,k}=NS_{i,k}+1$ or S: $m_j\neq\emptyset$, $(i \in I_{BBS-SO} \wedge n_i<K_i)$ S': Update (j,z) if $i \neq z$ then $\{\ n'_i=n_i+1,\ n'_z=n_z-1$ $\}$ $NS'_{i,k}=NS_{i,k}+1$

Table 4.7.e. Generator matrix Q_{Het}: diagonal entries

q(S,S')	relation between S and S'
$-\sum\limits_{S'' \in E_{Het},\ S''\neq S} q(S,S'')$	S' = S

service in the blocked node. In a network with BAS mechanism we observe that service blocking occurs after the completion of a job service while with BBS blocking occurs before the service. Hence, informally, the BAS mechanism produces less blocking than BBS. Finally, the BBS-O is the most blocking mechanism because all upstream nodes of a saturated node are blocked as soon as the node becomes full, even if the jobs in service are not destined to the full node.

For the two blocking types Stop and Recirculate the entire network or a given subnetwork satisfies some population constraints. Similarly to RS mechanism Recirculate blocking does not block the servers' activity while Stop mechanism blocks all the servers when the (sub)network reaches the minimum population threshold.

The exact performance analysis of the queueing networks with blocking based on the associated Markov process introduced in the previous section leads to the evaluation of the joint queue length distribution of the nodes in the networks in steady-state conditions and their average performance indices.

An exact solution algorithm of queueing networks with finite capacity and blocking based on the Markov process approach can be summarised as shown in Table 4.8.

Table 4.8. Solution algorithm for Markovian queueing networks with blocking

1	Definition of the system state space E.
2	Definition of the transition rate matrix Q that describes the queueing network evolution, according to the blocking type of each node.
3	Solution of linear system (4.4) to derive the stationary state distribution π at arbitrary times.
4	Computation from the solution vector π of the joint and marginal queue length distributions and of the average performance indices, such as throughput, utilization and mean response time, for each resource i of the network, $1 \leq i \leq M$.

The first two steps have been described in Section 4.1, where the state and the state space E have been defined for each type of blocking and the transition rate matrix Q has been defined by Tables 4.1 through 4.7.

The third step consists of the solution of the homogeneous linear system (4.4) subject to the normalization constant on state probabilities. Direct or iterative methods can be applied. Since the state space grows exponentially with the number of components of the network, efficient solution techniques have to be applied to solve this linear system.

It is easy to verify that the process matrix Q is always very sparse. This property grows with state space E cardinality. Hence one should use efficient storage methods for the system matrix, i.e. to store only the non-zero elements of Q, and solve the system with an iterative approach. However, by using iterative methods to solve the linear system convergence problems may be observed. This is a quite common problem in the analysis of queueing networks based on the Markov process analysis.

Moreover, note that state space E is finite for closed and open networks where all the nodes have finite capacity, but its cardinality grows exponentially with the number of nodes, the number of chains and the node population. The number of equations of system (4.4) is equal to the state space cardinality. For open networks, which include at least one infinite capacity queue, state space E is infinite and the solution π of the linear system (4.4) has to be approximated numerically.

Under certain constraints, depending both on the network definition and the blocking mechanism, special solution for π can be defined. In particular in some cases π has a product form solution and efficient solution algorithms can be applied to derive the performance indices. Hence, the algorithm above can be substituted by the direct evaluation of the closed form solution for which computationally efficient exact solution algorithms can be defined. This class of models will be described in chapter 5.

In the last step of the algorithm for the exact analysis of queueing networks with blocking we compute the performance indices from the solution of the linear system. Such solution of the global balance equations defines vector π that may directly correspond to the joint queue length distribution of the network, depending on the state definition. However, in the general case the state definition includes the queue length as one component together with other information that allow defining a Markov process. Hence the joint queue length distribution and the node queue length distribution can be obtained by summing the appropriate subset of state probabilities, depending on the blocking type.

Average performance indices include node and network utilization, throughput, mean queue length and mean response time. They are obtained by the stationary state joint distribution as shown in Chapter 2, Section 2.6. More precisely these performance indices have been defined in Section 2.6 as functions of network parameters and of

- $\pi_i(n_i)$, the stationary (marginal) queue length distribution of node i, where n_i is the number of customers in node i, $n_i \geq 0$, $1 \leq i \leq M$
- $\zeta_i(n_i, z_i)$, the stationary joint distribution of variables n_i and z_i, where z_i is the number of active servers at node i (not empty and not blocked servers), $0 \leq z_i \leq \min\{n_i, K_i\}$, where K_i is the number of servers of the node.

These probabilities depend on the blocking type X as follows:

$$\pi_i(n) = \sum_{S \in E_X : \pi-\text{cond}(X,S,n)} \pi_X(S)$$

$$\zeta_i(n,z) = \sum_{S \in E_X : \zeta-\text{cond}(X,S,n,z)} \pi_X(S)$$

where the appropriate subsets of probability π_X are identified respectively by the constraint π-cond(X, **S**, n) and ζ-cond(X, **S**, n, z), for each blocking type X. These conditions are defined in Table 4.9.

Table 4.9. Constraints to compute marginal probability $\pi_i(n)$ and $\zeta_i(n,z)$

Blocking Type	π-cond(X, **S**, n)	ζ-cond(X, **S**, n, z)
RS-RD and RECIRC.	$S_i = n_i = n$	$S_i = n_i = n$, $z = \min \{n, K_i\}$
BAS	$S_i = (n_i, s_i, \mathbf{m}_i)$, $n_i = n$	$S_i = (n_i, s_i, \mathbf{m}_i)$, $n_i = n$, $s_i = K_i - z$, $z = \min \{n, K_i\}$
BBS-SO BBS-SNO	(Blocked(i) $\wedge S_i = (n_i, NS_i)$, $n_i = n$) $\vee$ ($\neg$ Blocked(i) $\wedge S_i = n_i = n$)	(Blocked(i) $\wedge S_i = (n_i, NS_i)$, $n_i = n$ $z = NS_{i,0} + \sum_{k \in Dest_i : n_k < B_k} NS_{i,k}$) $\vee$ ($\neg$ Blocked(i) $\wedge S_i = n_i = n$, $z = \min \{n, K_i\}$)
BBS-O	$S_i = n_i = n$	$S_i = n_i = n \wedge \beta(S) = 1$, $z = \min \{n, K_i\}$
STOP	$S_i = n_i = n$	$S_i = n_i = n \wedge z = \min \{n, K_i\}$, $\Sigma_{1 \leq j \leq M} \, n_j > L$

4.3 ARRIVAL TIME DISTRIBUTION

The stationary joint queue length distribution at arrival times at a given node is a performance index of interest and it is used for exampleto compute the stationary waiting time distribution.

The arrival theorem for product-form queueing networks with infinite capacity queues relates the arrival time distribution with the queue length distribution at arbitrary times. Informally, the arrival theorem states that a customer arriving at a node of an open network sees the stationary state distribution of the network, and a customer arriving at a node of a closed network sees the stationary state distribution of the network with one customer less. The arrival theorem can be extended to some cases of product form networks with blocking, but with a different interpretation. Under certain constraints the arrival theorem still holds for product form queueing networks with some blocking types including Stop and Recirculate blocking and one can derive a necessary and sufficient condition under this product form condition. The validity of the arrival theorem for non product form networks with BAS, BBS and RS blocking types queueing networks is based on the analysis of the arrival time distribution and the result can be applied to some product form queueing networks.

In chapter 5 we present some results on the arrival theorem for product form queueing networks with blocking.

Informally, the arrival time distribution in queueing networks with blocking is similar to the stationary state distribution, but depends on the state of the two nodes between which the customer is moving.

The evaluation of the stationary state distribution is based on the definition of the Markov process underlying the queueing network introduced in Section 4.1. Let M denote this process. In order to evaluate the arrival time distribution we define a new homogeneous discrete time Markov process M^e embedded in process M. Let A denote the discrete state space of M^e and S^e the system state as seen by an arriving job at input time at node i, where the state does not include the arriving job. Informally, each state S^e of the embedded process M^e is identical, except for one less job at node i, to a corresponding state of the process M, just after the customer transition to node i, denoted by S^a. As in process M, also the state space definition of process M^e depends on the blocking type.

If the embedded Markov chain M^e is irreducible and recurrent then there exists the stationary state distribution ξ at arrival instants at node i. The direct evaluation of this distribution is not trivial. However, for a class of networks with finite capacity one can derive an expression of ξ in terms of the stationary state distribution at arbitrary times, π. By applying this result to some special cases of product-form networks with blocking, an extension of the arrival theorem for queueing networks with infinite capacity queues to networks with finite capacity queues and blocking can be derived.

For a Markovian non product form networks the following relationship between stationary distributions ξ and π holds. The following theorem can be proved for closed exponential networks with a general routing topology and blocking types BAS, BBS-SO and RS-RD

Theorem 4.1
The stationary state probability distributions ξ and π of a closed exponential network with finite capacity queues and blocking of type BAS, BBS-SO or RS-RD, are related as follows:

$$\xi(S^e) = \frac{1}{\eta} \sum_{S \in I(S^a)} \pi(S)q(S,S^a)$$

where $S^e \in A$, $S^a \in E$ is the state corresponding to S^e and, according to the blocking type:

$I(S^a)$ is the set of initial states of process M which occur just before a customer transition which leads to state S^a,

$q(S,S^a)$ is the transition rate from state S to S^a of process M and

η is a normalising constant.

We omit the proof of the theorem and the detailed definition of set $I(S^a)$ and transition rates $q(S,S^a)$ that can be found in the references. When the finite capacity node i has only one upstream node, say k (i.e., $p_{ki}>0$ and $p_{ji}=0$, $j\neq k$, $1\leq j\leq M$) then the set $I(S^a)$ is formed by a single state $S=S^a+ e_k - e_i$ and, for blocking type RS and BBS-SO, this relationship can be simplified as follows:

Corollary 4.1
If the node has only one upstream node, then

$$\xi(S^e) = \frac{1}{\eta} \pi(S) \qquad (4.9)$$

where $S^e \in A$, $I(S^a)=\{S\}$ and η is a normalising constant.

This result is an extension of a similar relationship for queueing networks with infinite capacity to networks with finite capacity queues. As a consequence, the evaluation of the steady-state probability distribution at arrival times ξ can be reduced to the evaluation of the probability distribution at arbitrary times π. This arrival time distribution can be used in the analysis of job passage time distributions in the network. Moreover the theorem can be simplified for a class of product-form networks with finite capacity, leading to an extension of the arrival theorem for queueing networks with infinite capacity queues to networks with finite capacity and certain blocking types.

Examples of networks that satisfy the condition of Corollary 4.1 are cyclic networks and central server networks where the central node has infinite capacity. This result allows an efficient computation of the passage and cycle time distribution.

4.4   CYCLE TIME DISTRIBUTION

The time spent by a customer in the system or in a subsystem (the passage time) is an important performance measure, which provides a more detailed performance evaluation of system behaviour than the average indices. The passage time distribution in queueing networks is generally difficult to obtain even for queueing networks with infinite capacity. For queueing networks with finite capacity queues and blocking, a few results have been obtained in terms of cycle time distribution for cyclic models and for central server model or star topology networks.

The evaluation of passage time distribution is based on the transient analysis of a stochastic process associated to the network and different from the Markov process introduced in Section 4.1 to evaluate the queue length distribution.

We can define a recursive algorithm to derive the cycle time distribution for cyclic closed exponential queueing networks with $M\geq 2$ finite capacity nodes and BBS-SO blocking and for M=2 nodes and BAS blocking. The method is based on the definition of a transient Markov process that describes the evolution of a specific (tagged) customer in a complete walk through the network. Sets E_0 and F of the possible initial and final states of the network are defined, corresponding to the beginning and the end of the walk of the tagged customer, respectively.

The cycle time distribution is computed by evaluating the first hitting time probability distribution of the Markov process to the final states, starting from the initial states.

Let Z denote a state of the transient Markov process and let T(Z,s) denote the Laplace-Stieltjes transform (LST) of the passage time distribution from Z to the final states F.

The LST of the cycle time distribution, denoted by T(s) can be computed as follows:

$$T(s) = \sum_{Z \in E_0} Prob(Z) \, T(Z,s) \qquad (4.10)$$

where $Prob(Z)$ is the probability of state Z at the cycle starting time and T(Z,s) can be computed by a recursive scheme. This recursive scheme can be reduced by taking into account the process structure and the blocking type definition. The reader interested in the complete definition of the algorithm may refer to the references given in Section 4.6.

Since each state Z corresponds to a system state $S^e \in A$ of the embedded process M^e, as introduced in the previous section to define the joint queue length distribution, then $Prob(Z)$ can be evaluated as the stationary distribution at arrival times, $\xi(S^e)$ and, by applying Theorem 4.1, as a function of the stationary distribution at arbitrary times, $\pi(S)$, $S \in E$.

From the recursive scheme to evaluate the LST of the cycle time distribution one can derive an explicit expression for the cycle time distribution in the time domain, where coefficients are defined by recursion.

For a two-node closed exponential network with BBS-SO or BAS blocking this approach leads to the following closed-form expression in the time domain of the density function $f(t)$ of the cycle time:

$$f(t) = \sum_{j=1}^{3} e^{-\mu_j t} \sum_{i=1}^{k_j} c_{ji} \frac{t^{i-1}}{(i-1)!} \qquad (4.11)$$

where $k_1 = k_2 = N$, $k_3 = 2N-3$, N customers in the network, $\mu_3 = \mu_1 + \mu_2$ and coefficients c_{ij} are recursively computed for each i and j. The evaluation of coefficients c_{ij} requires a computational complexity of $O(N^3)$.

Note that for the special case of a two-node network with blocking, the stationary distribution π has a product-form solution and consequently ξ and $Prob(Z)$ have a product form solution. In this case an extension of the arrival theorem holds, as discussed in Section 4.3.

However, the algorithm sketched above to evaluate the cycle time distribution applies to any Markovian non product-form network. For the class of product-form networks with finite capacity, we take advantage of a more efficient computation of distribution π and consequently of $Prob(Z)$ in formula (4.10).

In many practical applications it is sufficient to evaluate the first few moments of the cycle time. A recursive evaluation of the cycle time moments can be derived for cyclic closed exponential networks and for central server model networks.

For a two node exponential network with BBS-SO or BAS blocking, the k-th moment of the cycle time distribution, $E(k)$, for $k=1,2,\ldots$, is given by:

$$E(k) = \sum_{j=1}^{3} \sum_{i=1}^{k_j+1} \frac{c_{ji}}{\mu_j^{i+k}} \frac{(i+k-1)!}{(i-1)!} \qquad (4.12)$$

where k_j, $1 \leq j \leq 3$, are defined as in formula (4.11).

The evaluation of the passage time distribution in other classes of queueing network models with finite capacity, including different types of blocking and non-exponential service time distribution is an open issue.

Example 4.5 Consider a two node cyclic network with $N=10$ customers, service rates $\mu_1=1$, $\mu_2=2$, BBS-SO blocking and finite capacities B_1 and B_2. Figure 4.10 illustrates the cycle time distribution f (t) by varying the queue capacities B_1 and B_2, where $B_1 + B_2 = B = 1,2,3,4,10$. Note that the cycle time distribution depends only on values N and B, while it is independent of the single finite capacities B_1 and B_2. That is the cycle time distribution of the two node cyclic network with finite capacities B_1 and B_2 is identical to the one the network where only one node has finite capacity equal to B.

We osbserve the impact of the finite capacity on the cycle time distribution. As factor B increases the cycle time distribution approaches the one with $B=N=10$ that correspond to the model with infinite capacity queues.

Such parametric analysis can be helpful in system design, by determining the minimum capacity B that guarantees a required performance quality, i.e. a required cycle time value. For example, from Figure 4.10 one can derive that the cycle time is guaranteed to be less than or equal to 14 with 0.75 probability for the minimum $B=2$ and for $B=3,\ldots,10$ with probability greater than or equal to 0.75.

4.5 BIBLIOGRAPHICAL NOTES

Stochastic processes and Markov processes as model of queueing networks are introduced in Kleinrock (1975). Baskett et al. (1975) define the Markov process for the class of product-form BCMP networks with infinite capacity queues. The definition of the Markov process underlying queueing networks with Stop and Recirculate blocking is given in Van Dijk (1991a) and (1991b).

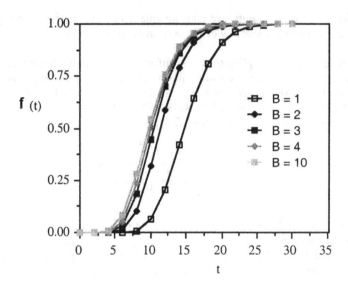

Figure 4.10. Cycle time distribution in a two node network with blocking and finite capacity
$$B = B_1 + B_2$$

The arrival theorem for product-form queueing networks with infinite capacity queues is proved in Lavenberg and Reiser (1980) and Sevcik and Mitrani (1981). Boucherie and Van Dijk (1997) extend the arrival theorem to some cases of product form networks with blocking, but with a different interpretation. They discuss the validity of the arrival theorem for product form networks with different blocking types including Stop and Recirculate blocking and derive a necessary and sufficient condition under this product form. The arrival theorem for special case of two-node product form exponential networks with BBS-SO or BAS blocking is proved in Balsamo and Donatiello (1989a) and (1989b).

The arrival theorem for non product form networks with BAS, BBS and RS blocking types queueing networks has been discussed in Balsamo and Clò (1992) and the result is applied to some product form queueing networks. The reader interested in the proof of theorem 4.1, corollary 4.1 and the detailed definition of set $I(S^a)$ and transition rates $q(S, S^a)$ may refer to this reference.

As concern sojourn time and passage time distribution, a survey of sojourn time results in queueing networks is presented in Boxma and Daduna (1990). The algorithms to evaluate the cycle time distribution and the cycle time moments for two node exponential closed queueing networks with finite capacity queues for BBS-SO and BAS blocking are given in Balsamo and Donatiello (1989a) and (1989b). In particular they provide the recursive computation of coefficients c_{ij} to evaluate the density function $f(t)$ of the cycle time defined by formula (4.11) and the cycle time moments defined by formula (4.12). The algorithm to compute the cycle time distribution for cyclic exponential networks with $M \geq 2$ nodes and BBS-SO blocking can be found in Balsamo, Clò and Donatiello (1993). Balsamo and Clò (1993) provide the evaluation of the cycle time moments for closed exponential network with central server or star topology and blocking.

REFERENCES

Balsamo, S., and C. Clò "State distribution at arrival times for closed queueing networks with blocking" Technical Report TR-35/92, Dept. of Comp. Sci., University of Pisa, 1992.

Balsamo, S., and C. Clò "Delay distribution in a central server model with blocking", Technical Report TR-14/93, Dipartimento di Informatica, University of Pisa, 1993.

Balsamo, S., C. Clò, and L. Donatiello "Cycle Time Distribution of Cyclic Queueing Network with Blocking" Performance Evaluation, Vol. 17 (1993) 159-168.

Balsamo, S., and L. Donatiello "On the Cycle Time Distribution in a Two-stage Queueing Network with Blocking" IEEE Transactions on Software Engineering, Vol. 13 (1989) 1206-1216.

Balsamo, S., and L. Donatiello "Two-stage Queueing Networks with Blocking: Cycle Time Distribution and Equivalence Properties", in *Modelling Techniques and Tools for Computer Performance Evaluation* (R. Puigjaner, D. Potier Eds.) Plenum Press, 1989.

Baskett, F., K.M. Chandy, R.R. Muntz, and G. Palacios "Open, closed, and mixed networks of queues with different classes of customers" J. of ACM, 22 (1975) 248-260.

Boucherie, R., and N. Van Dijk "On the arrival theorem for product form queueing networks with blocking" Performance Evaluation, 29 (1997) 155-176.

Boxma, O., and H. Daduna "Sojourn time distribution in queueing networks" in 'Stochastic Analysis of computer and Communication Systems' (H. Takagi Ed.) North Holland, 1990.

Kleinrock, L., *Queueing Systems.Vol.1: Theory.* Wiley (1975).

Lavenberg, S. S., and M. Reiser "Stationary State Probabilities at Arrival Instants for Closed Queueing Networks with multiple Types of Customers" J. Appl. Prob., Vol. 17 (1980) 1048-1061.

Sevcik, K. S., and I. Mitrani "The Distribution of Queueing Network States at Input and Output Instants" J. of ACM, Vol. 28 (1981) 358-371.

Van Dijk, N. "On 'stop = repeat' servicing for non-exponential queueing networks with blocking" J. Appl. Prob., Vol. 28 (1991) 159-173.

Van Dijk, N. "'Stop = recirculate' for exponential product form queueing networks with departure blocking" Oper. Res. Lett., Vol. 10 (1991) 343-351.

5 EXACT ANALYSIS OF SPECIAL NETWORKS

In this chapter we deal with some special classes of queueing networks with blocking for which efficient solution algorithms can be defined. The exact analysis of Markovian networks based on the Markov process underlying the network defined in Chapter 4 can be dramatically simplified for these particular networks.

Specifically, in Section 5.1 we introduce the class of networks with product form solutions whereas the convolution algorithm to evaluate performance indices is described in Section 5.2.

Section 5.3 deals with the class of symmetrical networks where all the nodes have the same finite capacities and the customers show the same behaviour. In this case we define an aggregation technique which leads to a drastically reduction of the state space of the Markov process associated to the network. Moreover, we can apply a further reduction of the process state space for product form symmetrical networks.

Section 5.4 presents various results concerning the arrival theorem for the class of product form networks with blocking.

5.1   PRODUCT FORM NETWORKS

In this section we introduce the class of queueing networks with finite capacity and blocking that have a product form solution. Since this class is a subset of Markovian networks with finite capacity considered in the previous chapter, the same analytical techniques can be applied to solve these networks.

However, an important consequence of the identification of the class of product form networks is that this property allows to define efficient algorithms to evaluate performance indices. Specifically, in some cases one can extend basic algorithms for the class of product form BCMP networks with infinite capacity queues, such as MVA and Convolution algorithms to queueing networks with finite capacity queues.

First we summarise the cases of product form networks with finite capacity and different blocking types. The extension of the arrival theorem to some cases of this class of networks is discussed in Section 5.4.

5.1.1 Product form solution of the joint queue length distribution

Product form solutions of the joint queue length distribution π for single class open or closed networks under certain constraints, depending both on the network definition and the blocking mechanism, can be defined as follows:

$$\pi(S) = \frac{1}{G} V(n) \prod_{i=1}^{M} g_i(n_i) \qquad (5.1)$$

where G is a normalising constant and n is the total network population. The functions V and g_i, $1 \leq i \leq M$, are defined in terms of network parameters which include vector **x** defined by system (2.1) of traffic balance equations introduced in Chapter 2 and service rates μ_i, $1 \leq i \leq M$, and depend on the blocking type and additional constraints.

Similarly, for multichain open, closed or mixed queueing networks with blocking, formed by M nodes and R chains, product form solutions can be defined as follows:

$$\pi(S) = \frac{1}{G} \prod_{r=1}^{R} V_r(N_r) \prod_{i=1}^{M} g_i(n_i)$$

where G is a normalising constant, N_r is the total network population in chain r, $1 \leq r \leq R$, and the functions V_r, $1 \leq r \leq R$, and g_i, $1 \leq i \leq M$, are defined in terms of network parameters.

For the sake of readability in the following we only present formulas of product form solutions for single class networks; the detailed expression of functions V_r, $1 \leq r \leq R$, and g_i, $1 \leq i \leq M$, in product form solutions for multiclass networks is given in the reference reported in Section 5.5.

Table 5.1 summarises product form networks with finite capacity and different blocking types.

Five network topologies are considered for both homogeneous networks, where each node works under the same blocking type, and non-homogeneous ones, where different nodes in the networks work under different blocking mechanisms.

The first three topologies concern closed networks and are the two-node network, the cyclic topology and the central server or star topology. For the central server topology networks, node 1 denotes the central node and the routing matrix P is defined as follows: $p_{ij} > 0$ for i=1, $2 \leq j \leq M$, $p_{i1} = 1$ for $2 \leq i \leq M$ and $p_{ij} = 0$ otherwise, $1 \leq i, j \leq M$.

The fourth case refers to queueing networks with reversible routing matrix P. A routing matrix P is said to be reversible if $x_i p_{ij} = x_j p_{ji}$, and $\lambda_0 p_{0j} = x_j p_{j0}$ for $1 \leq i, j \leq M$.

Table 5.1 shows the cases of product form solution together with additional constraints, for each combination of blocking type and network topology, where:
- PFi denotes the corresponding product form formula, $1 \leq i \leq 8$, defined below for i=3 and 6 and in Table 5.2 for all the other cases;
- an arrow denotes that the case is included in the more general class of arbitrary topology networks and, as far as we are aware, there are no special results which only hold for that specific topology;
- 'NO' means that, as far as we are aware, no product form has been proved;
- 'NA' means that the blocking type is not applicable to the network topology;
- for non-homogeneous networks the allowed combination of blocking types is also given.

Some additional conditions are required in some cases.

Let $B = \sum_{i=1}^{M} B_i$ denote the total capacity of the network and let $B_{min} = min \{B_j, 1 \leq j \leq M\}$.

Definition 5.1
The **non-empty condition** for closed networks requires that at most one node can be empty, i.e., $N \geq B - B_{min}$.

This condition is said to be **strictly** verified when each node can never be empty, i.e., if the inequality strictly holds.

Definition 5.2
The condition which requires **at most one blocked node** is satisfied if $N = B_{min} + 1$.

In other words, if a node is full then at most one of its sending nodes is not empty and can be blocked.

Definition 5.3
Condition (A) refers to a particular model of multiclass networks with parallel queues with interdependent blocking functions and service rates, and which satisfy a so-called invariant condition. See references in Section 5.5 for further details.

Definition 5.4
Condition (B) requires that each node i with finite capacity is the only destination node for each upstream node, i.e., it satisfies the following constraint:

if $p_{ji} > 0$ then $p_{ji} = 1$, $1 \leq j \leq M$.

Table 5.2 shows the definitions of product forms PFi, for i=1,2,4,5,7 and 8, in terms of conditions on the network model and expressions for functions V and g_i, $1 \leq i \leq M$, in product form (5.1) for single class networks. In Table 5.2, for product form PF4, A-type nodes are defined as follows:

Table 5.1. Product form networks with blocking

Blocking Type	Two-Node	Cyclic	Central Server	Reversible Routing	Arbitrary
		network	topology		
BAS	PF1	$\rightarrow$	$\rightarrow$	$\rightarrow$	PF7 at most one blocked node
BBS-SNO	PF1 if $N \leq B_1 + B_2 - 2$	NO	NO	NO	NO
BBS-SO	PF1	PF2 non-empty condition	PF3 if only $B_1 < \infty$	$\rightarrow$	PF2 strictly non-empty condition and cond. (B)
RS-RD	PF1	PF2 non-empty condition	PF3	PF4 PF6 and cond. (A)	PF2 strictly non-empty condition
RS-FD	PF1	PF2 non-empty condition	PF3 if only $B_1 < \infty$	$\rightarrow$	PF2 strictly non-empty condition and cond. (B)
Stop	PF5	NO	PF5	PF5	PF8
Recirculate	NA	NA	NA	$\rightarrow$	PF8
Non Homogeneous	BAS BBS-SO RS-RD RS-FD	BBS-SO RS-RD RS-FD non-empty condition	BBS-SO RS-RD RS-FD node 1 with RS	RS-RD Stop	BBS-SO RS-RD RS-FD strictly non-empty condition and cond. (B) for BBS-SO and RS-FD
	PF1	PF2	PF3	PF5	PF2

Definition 5.5

A **node** is said to be **A-type** if it has an arbitrary service time distribution and a symmetric scheduling discipline or exponential service time distribution, which is the same for each class at the same node, when the scheduling is arbitrary.

We shall now define the product form solutions PF3 and PF6.

PF3:
Conditions:
- multiclass central server networks with the class type of a job fixed in the system
- state-dependent routing depending on the class type
- blocking functions dependent on node and class
- A-type nodes.

Table 5.2. Product form formulas and conditions

	Conditions	V(n)	$g_i(n_i), \forall n_i, \ 1 \leq i \leq M$
PF1	multiclass networks BCMP type nodes class independent capacities	1	$(x_i / \mu_i)^{n_i}$
PF5	like PF4, but single class		
PF7	multiclass networks FCFS-exponential nodes class independent capacities		
PF8	multiclass open Jackson networks with class type fixed		
PF4	multiclass networks with class type fixed blocking functions dep. on node, class and chain A-type nodes load dependent service rates $\mu_i(n_i) = \mu_i f_i(n_i)$	1	$(x_i / \mu_i)^{n_i} \prod_{l=1}^{n_i} \dfrac{b_i(l-1)}{f_i(l)}$
PF2	single class networks exponential nodes load independent service rates with $\varepsilon = (\varepsilon_1, \ldots, \varepsilon_M)$ $\varepsilon = \varepsilon \, \mathbf{P'}$ $\mathbf{P'} = \| p'_{ij} \|, \ p'_{ij} = \mu_j p_{ji}, \ i \neq j,$ $p'_{ii} = 1 - \Sigma_{j \neq i} p'_{ji}, \ 1 \leq i, j \leq M$	1	$1 / \varepsilon_i^{n_i}$

For single class exponential networks with load dependent service rates $\mu_i(n_i) = \mu_i f_i(n_i)$ and state-dependent routing $p_{1j}(n_j) = w_j(n_j) \, w(N-n_1) \ \forall n_j, \ p_{j1} = 1$ for $2 \leq j \leq M$, where M is node number and N customer number in the network, product form (5.1) holds with

$$V(N) = \prod_{l=1}^{N-n_1} w(l-1) \prod_{j=2}^{M} \prod_{l=1}^{n_j} w_j(l-1), \ g_i(n_i) = \prod_{l=1}^{n_i} \frac{1}{\mu_i} \frac{b_i(l-1)}{f_i(l)} \ , \forall n_i, \ 1 \leq i \leq M$$

For a complete definition see the references.

PF6:
Conditions:
- multiclass networks with the class type of a job fixed in the system
- interdependent blocking probability and service rates

• A-type nodes.

For single class networks, product form (5.1) holds with

$$V(n) = 1, n{\geq}0, g_i(n_i) = (x_i)^{n_i} h_i (\mu_i, n_i), n_i{\geq}0, 1{\leq}i{\leq}M$$

where $h_i (\mu_i, n_i)$ is a product form function dependent on the state and the service rate of node i and is defined according to the scheduling of the interdependent parallel queues. For a complete definition see the references reported in Section 5.5.

The proof that product form (5.1) is a solution of the Markov process associated to the network can be obtained by substituting the closed-form expression into the global balance equations of the underlying Markov process (linear system (4.4) in Chapter 4, Section 4.1).

Note that product form expression (5.1) generalizes the closed-form expression for BCMP networks, and in certain cases, such as PFi, i=1,5,7 and 8, corresponds to the same solution as for queueing networks with infinite capacity queues computed on the truncated state space of the network with finite capacities.
This relationship provides the basis for the equivalence between product form networks with and without blocking that will be presented in Chapter 7, Section 7.1.2.

Observations
Identification of necessary and sufficient conditions under which a queueing network model has a product form solution is an open issue for networks with infinite capacity queues as well.
We observe that most of the product form solutions for queueing networks with finite capacity queues have been derived by using the following properties:
• reversibility of the underlying Markov process,
• duality.

The first approach can be applied to networks with finite capacity whose underlying Markov process is shown to be obtained by truncating the reversible Markov process of the network with infinite capacity. Hence, a product form solution immediately follows from the theorem for truncated Markov processes of reversible Markov processes. This theorem states that the truncated process shows the same equilibrium distribution as the whole process normalised on the truncated sub-space.

A product form solution of both homogeneous and non-homogeneous two-node cyclic networks can be proved by using this property for exponential single class networks and for multiclass networks with BCMP nodes under additional constraints.
Similarly, it has been proved that closed queueing networks with a reversible routing matrix P have a reversible underlying Markov process under RS-RD or Stop blocking and different types of nodes This class of product form networks with RS-RD blocking for multiclass networks has been extended to include A-type nodes and

more general blocking functions which may depend both on the total population, class population and routing chain population at the node.

The central server or star topology network is a special case of product form networks with reversible routing. However, some product form results have only been specifically proved for central server networks.

Remark. Although some of these results concern networks where routing probabilities are dependent on the state of the network, they are related to queueing networks with finite capacity and blocking, as discussed in Section 2.3 of Chapter 2. Indeed, by using blocking functions, the actual routing probabilities of queueing networks with finite capacity can be interpreted as state dependent probabilities, and they are obtained by combining the routing probability matrix P with blocking functions $b_i(n_i)$, $1 \leq i \leq M$. Therefore, product form solutions for networks with state dependent routing can be interpreted in the same way as for blocking networks, and can be extended to multiclass networks. A generalization of these results can be obtained by combining state dependent routing and finite capacity queues.

Routing reversibility, which leads to the Markov process reversibility is related to the so-called job-local-balance of the underlying Markov process. This balance property, which is related to local balance and station balance for queueing networks with infinite capacity, states that the rate outside a state due to any particular job in the system is equal to the rate inside that state which is due to that particular job. Job-local-balance provides the basis for deriving equivalence and insensitivity properties that we shall present in Chapter 7.

The second approach to derive product form solution of queueing networks with blocking is based on duality. Product form PF2 has been obtained by adding the capacity constraint to the Gordon-Newell closed exponential networks and by defining a dual network which has the same stationary joint queue length distribution.

Consider a cyclic closed network with M nodes, N customers, node capacities B_i, $1 \leq i \leq M$ and BBS-SO or RS blocking. The dual network is obtained from the original one by reversing the connections between the nodes. It is formed by M centers and (B - N) customers which correspond to the 'holes' of the original (primal) network, where B is the total capacity of the network defined above. When a customer moves from node i in the original network, a hole moves backward to node i in the dual one. When there are n_i customers in node i of the original network, the i-th center of the dual one contains $B_i - n_i$ holes, $1 \leq i \leq M$. It can be shown that the underlying Markov process which describes the evolution of customers in the network is equivalent to the one which describes the behaviour of holes in the dual network. As a consequence, when the non-empty condition is satisfied, then the total number of holes in the dual network cannot exceed the minimum capacity, i.e., (B - N)$\leq B_{min}$, and the dual network has a product form solution like a network without blocking. Hence the product form solution for the primal network is given by formula (5.1) with V(N)=1 and $g_i(n_i) = (1/\mu_{i-1})^{n_i}$, $1 \leq i \leq M$, (where if i=1 then i-1=M), which corresponds to expression PF2. This solution can be extended to arbitrary topology networks with load independent service rates for RS-RD blocking. Moreover, this result can be further extended to homogeneous networks with BBS-SO blocking under condition (B) (note that as we shall see in chapter 7 in this case blocking BBS-SO and BBS-O are identical) and to heterogeneous networks.

The concept of duality has been applied to exponential closed cyclic networks with RS-RD blocking and to cyclic networks with phase-type service distributions and BBS-SO blocking. In both cases the throughput of the network is shown to be symmetric with respect to its population, as we shall discuss in Section 5.3.

5.2   ALGORITHMS   FOR   PRODUCT   FORM NETWORKS

Product form solution (5.1) allows us to define efficient algorithm to evaluate performance indices of networks with blocking. The definition of the average performance indices and the queue length distribution of product form networks with blocking can be expressed similarly to that of product form networks without blocking from the state probability. However, for networks with blocking the finite capacity queues limit the state space of the network and the definition of the queue length distribution and the average indices depends on the blocking type, as defined in Chapter 4. Therefore we cannot immediately apply the known algorithms already known for BCMP networks, such as convolution algorithm and MVA. The state space limitation leads to a new definition of a convolution algorithm taking into account a set of constraints on the queue lengths.

An MVA algorithm can be defined for the class of product form networks with cyclic topology and with BBS-SO and RS blocking. In this case product form PF2 solution holds when the non-empty condition is satisfied and we can define a dual network without blocking with identical product form state distribution, as described in Section 5.1.1. Then by using the duality property this MVA algorithm simply applies the standard MVA algorithm for networks without blocking to the dual network (for details see the references). Note that this MVA algorithm is not related to the arrival theorem as the MVA for queueing networks without blocking, since it is based on the dual network that is without blocking. The arrival theorem for networks with blocking is discussed in Section 4.3 and for product form networks with blocking in Section 5.4.
Another MVA algorithm has been recently extended to a class of product form networks and with RS blocking, based on recursive relations for average performance indices on the network with blocking. It applies to networks with PF2 or PF3 product form solution and has a time computational complexity of $O(\max\{B_i\} \, M \, N)$ operations.

We shall now introduce a convolution algorithm for product form queueing networks with blocking, whose computational complexity is polynomial in the number of network components. Specifically, the convolution algorithm has a linear time computational complexity in the number of network components, that is it requires $O(M \, N)$ operations.

5.2.1 Convolution algorithm

Consider single class networks with load independent exponential service centers and with single type of customers.

We present a convolution algorithm for product form queueing networks with blocking. This algorithm has a time computational complexity linear in the number of network components, i.e., the number of service centers and the number of customers.

The algorithm is based on a set of recursive equations used to compute the normalizing constant G in formula (5.1), from which the marginal queue length distribution of each service center and the average performance indices are derived by closed-form expressions.

Let $a_i = \max(0, N - \sum_{1 \leq j \leq M, j \neq i} B_j)$ denote the minimum feasible queue length of service center i in the network, $1 \leq i \leq M$, i.e., the number of customers in service center i satisfies the following constraints: $a_i \leq n_i \leq B_i$. In order to define the convolution algorithm we consider the following two definitions of function g_i which hold for various blocking types and under different constraints:

$$g_i(n_i) = \rho_i^{n_i}, \quad n_i > 0, \quad 1 \leq i \leq M \qquad (5.2)$$

with
- case I: $\rho_i = x_i / \mu_i$
- case II: $\rho_i = 1 / \varepsilon_i$

and where ε has been defined in Table 5.2 for the product form PF2.
These expressions hold as illustrated in Table 5.2.
In particular case I holds for
- a network with RS-RD or Stop blocking, reversible routing, nodes with arbitrary scheduling
- a two-node network with blocking types BAS, BBS, RS and Stop, and FIFO service discipline. Note that in this case we use BBS and RS to denote the general type of blocking because one can show that two networks with BBS-SO, BBS-SNO and BBS-O blocking have the same behavior as well as two networks with RS-RS and RS-FD; these and other equivalences are presented in Chapter 7.

Case II holds for
- a network with arbitrary topology, RS-RD blocking and where the number of customers N verifies the strictly non-empty condition defined by Definition 5.1
- networks with RS-RD or RS-FD or BBS-SO blocking, when the strictly non-empty condition holds and each node i with finite capacity has a single sending node j, i.e., if $p_{ij} > 0$ then $p_{ij} = 1$, $1 \leq j \leq M$. Note that in this case one can show that two networks with RS-RS and RS-FD blocking have the same behavior as we shall present in Chapter 7.

- a network with cyclic topology, BBS-SO or RS blocking and where the non-empty condition holds. Note that in this case RS-RD and RS-FD have the same behavior (see Chapter 7).

The convolution algorithm computes the normalising constant G in formula (5.1). Let us define function $G_j(n)$ as follows:

$$G_j(n) = \sum_{(n_1,\dots,n_j)\in E(j,n)} \prod_{1\le i\le j} g_i(n_i) \qquad (5.3)$$

where

$$E(j,n) = \{(n_1,\dots,n_j): a_i \le n_i \le B_i, \ 1\le i \le j, \ \sum_{1\le i\le j} n_i = n\}$$

for $A_j \le n \le B(j)$, $1\le j\le M$, where $A_j = \sum_{1\le i\le j} a_i$, $B(j) = \sum_{1\le i\le j} B_i$ and a_i and B_i are the minimum and maximum queue lengths of node i.

$E(j,n)$ is the set of state of the network formed by the first j nodes and with n customers, but with the population constraints referred to the overall network, i.e. the minimum feasible queue length of center i a_i is computed by considering all the nodes.

For queueing networks with infinite capacity queues $G_j(n)$ is the normalizing constant of the network with the first j nodes and n customers, $1\le j\le M$ and $0\le n\le N$. However, note that this is not always true for networks with finite capacity when $j<M$ and $n<N$. Indeed, the constraint $n_i\ge a_i$ in formula (5.3) depends on the network parameters N and B_j, $1\le j\le M$. $G_j(n)$ is the normalising constant of the network with the first j nodes and n customers only when $a_i=0$, $1\le i\le j$. However, hereafter, we use this interpretation of function $G_j(n)$ in any case, for the sake of simplicity.

By definition, the overall normalising constant is $G = G_M(N)$. To evaluate G we use the following recursive equations to compute functions $G_j(n)$, $A_j\le n\le B(j)$, $1\le j\le M$.

Theorem 5.1
For $n\le N$ and $2\le j\le M$,
i) if $A_{j-1}+B_j > B(j-1)+a_j$ then

$$G_j(n)= \rho_j^{a_j} \, G_{j-1}(A_{j-1}) \qquad \text{if } n = A_j \qquad (5.4)$$

$$G_j(n)= \rho_j^{a_j} G_{j-1}(n-a_j)+\rho_j \, G_j(n-1) \qquad \text{if } A_j+1\le n\le B(j-1) + a_j \quad (5.5)$$

$$G_j(n) = \rho_j G_j(n-1) \qquad \text{if } B(j-1) + a_j + 1 \leq n \leq A_{j-1} + B_j \qquad (5.6)$$

$$G_j(n) = -\rho_j^{B_j+1} G_{j-1}(n-B_j-1) + \rho_j G_j(n-1) \qquad (5.7)$$
$$\text{if } A_{j-1} + B_j + 1 \leq n \leq B(j) - 1$$

$$G_j(n) = \rho_j^{B_j} G_{j-1}(B(j-1)) \qquad \text{if } n = B(j) \qquad (5.8)$$

ii) if $A_{j-1}+B_j \leq B(j-1)+a_j$ then

if $n = A_j$, $G_j(n)$ is given by formula (5.4);
if $A_j+1 \leq n \leq A_{j-1} + B_j$, $G_j(n)$ is given by formula (5.5);
if $A_{j-1} + B_j + 1 \leq n \leq B(j-1) + a_j$, then

$$G_j(n) = \rho_j^{a_j} G_{j-1}(n-a_j) - \rho_j^{B_j+1} G_{j-1}(n-B_j-1) + \rho_j G_j(n-1) \qquad (5.9)$$

if $B(j-1) + a_j + 1 \leq n \leq B(j) - 1$, $G_j(n)$ is given by formula (5.7);
if $n = B(j)$, $G_j(n)$ is given by formula (5.8);

and with initial condition, for $j = 1$

$$G_1(n) = \rho_1^n \qquad \text{for } a_1 \leq n \leq B_1 = B(1) \qquad (5.10)$$

Proof. The initial condition given by formula (5.10) derives from formula (5.3). To derive the set of recursive equations, we consider the generating function $G_j(z)$ of $G_j(n)$ and two different computational approaches.

By the node generating function we can define the generating function $G_j(z)$ related to the first j nodes, $1 \leq j \leq M$ as follows:

$$G_j(z) = \frac{(\rho_j z)^{a_j} - (\rho_j z)^{B_j+1}}{1 - \rho_j z} G_{j-1}(z)$$

from which

$$G_j(z) = (\rho_j z)^{a_j} G_{j-1}(z) - (\rho_j z)^{B_j+1} G_{j-1}(z) + \rho_j z G_j(z) \qquad (5.11)$$

On the other hand, we can also write the generating function $G_j(z)$ as follows:

$$G_j(z) = \sum_{A_j \leq n \leq B(j)} G_j(n) z^n \qquad (5.12)$$

where function $G_j(n)$ is given by formula (5.3). By substituting expression (5.12) in equation (5.11) and some algebra, one obtains the following expression:

$$\sum_{A_j \le n \le B(j)} G_j(n)z^n = \rho_j^{a_j} \sum_{m=A_j}^{B(j-1)+a_j} G_{j-1}(m-a_j)z^m$$

$$-\rho_j^{B_j+1} \sum_{m=A_{j-1}+B_j+1}^{B(j)-1} G_{j-1}(m-B_j-1)z^m + \rho_j \sum_{m=A_j+1}^{B(j)+1} G_j(m-1)z^m$$

We equate the coefficients of the two polynomials in z of the two sides of the equation by considering whether the following condition holds: $A_{j-1}+B_j > B(j-1)+a_j$.

For example, the coefficients of z^n for $n = A_j$ are $G_j(A_j)$ and $\rho_j^{a_j}G_{j-1}(A_{j-1})$ on the left and the right hand sides of the equation, respectively.

By equating the two coefficients we immediately obtain the recursive equation (5.4). Likewise for $A_j < n \le B(j)$ we obtain the equations from (5.5) to (5.9). **QED**

If we compare the convolution algorithm for queueing networks with blocking with that for queueing networks with infinite capacity queues we note that the latter is based on the following recursive equation for load independent service rates:

$$G_j(n) = G_{j-1}(n) + \rho_j G_j(n-1), \text{ for } 1 \le j \le M, 0 \le n \le N. \quad (5.13)$$

This equation can be interpreted by considering that the total number of customers n in the network with the first j nodes can be distributed to the nodes as follows:

(a) zero customers in node j and n customers in the first j-1 nodes,

(b) at least one customer in node j and the remaining n-1 customers in the first j nodes (node j included).

Case (a) corresponds to the first term and case (b) to the second term of the right hand side of equation (5.13).

Function $G_j(n)$ defined by formula (5.3) for queueing networks with finite capacity queues can be interpreted as the normalizing constant of the network with the first j nodes and n customers. Note that in the special case of $a_j=0$ equation (5.5) reduces to equation (5.13). In the general case ($a_j \ge 0$) the proposed convolution algorithm evaluates function $G_j(n)$ for a range of values of the number of customers depending on node j, $A_j \le n \le B(j)$. Note that, except for the first node, the upper bound can be greater than the total network population, i.e., it can be $B(j) > N$. So the evaluation of $G_j(n)$ for $n > N$ is not necessary.

Figure 5.1 shows the range of values n of each function $G_j(n)$, $1 \le j \le M$. Figures 5.2 and 5.3 illustrate the recursive equations (5.4)-(5.9) for function $G_j()$ for the two cases (i) and (ii), respectively.

Equations (5.4) and (5.8) can be easily interpreted by noting that the number of customers $n=A_j$ and B(j) determine a unique distribution of the customers on the nodes. In the two cases, the only feasible state is $(n_1, \ldots, n_j)$ with components $n_i = a_i$ and $n_i = B_i$, $1 \le i \le j$, respectively. Therefore these two equations relate the normalizing constant of the network with the first j nodes and n customers with the constant $G_{j-1}(n-n_j)$. These expressions can be also directly obtained by formula (5.3).

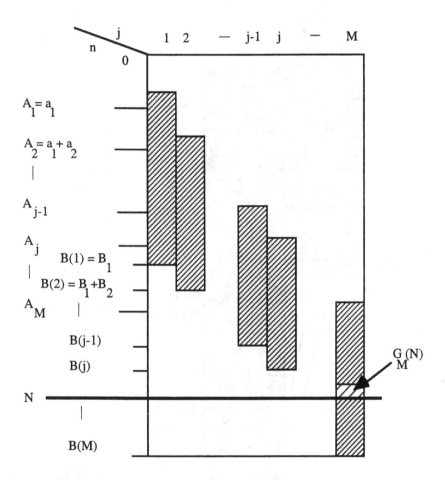

Figure 5.1. Scheme of the range n for each function $G_j(n)$, $A_j \le n \le B(j)$, $1 \le j \le M$, to evaluate the normalizing constant $G_M(N)$

Equation (5.5) is similar to equation (5.13) for infinite capacity queues. It relates the normalizing constant of the network with the first j nodes and n customers with $G_{j-1}(n-a_j)$ and $G_j(n-1)$, corresponding to the two following cases:
(a) a_j customers in node j, which is its minimum queue length, and the remaining $n-a_j$ customers in the first $j-1$ nodes,
(b) at least one customer in node j and the remaining $n-1$ customers in the first j nodes (node j included).

Case (a) and case (b) correspond respectively to the first and to the second term of the right hand side of equation (5.5).

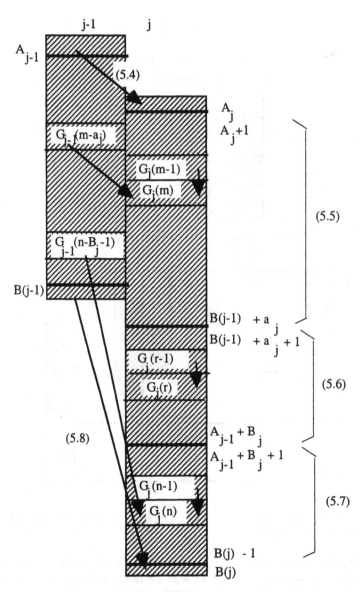

Figure 5.2. Scheme of recursive equations (5.4)-(5.8) for function $G_j(n)$, $A_j \leq n \leq B(j)$, $1 \leq j \leq M$, for case (i)

Equation (5.6) has the same meaning of equation (5.5), but without the term corresponding to case (a). Indeed, the number of customers is $n > a_j + B(j-1)$ and such that node j cannot attain its minimum queue length a_j even if all the first j-1 nodes have the maximum queue length (i.e., with $B(j-1)$ customers in the first j-1 nodes).

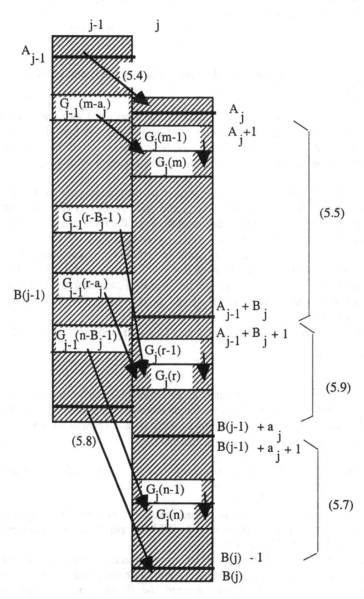

Figure 5.3. Scheme of recursive equations (5.4), (5.5) and (5.7)-(5.9) for function $G_j(n)$, $A_j \le n \le B(j)$, $1 \le j \le M$, for case (ii)

Equation (5.7) applies when the number of customers exceed the minimum overall capacity of the first j nodes (A_{j-1}) and node j capacity (B_j), i.e., for $n > A_{j-1} + B_j$. This equation is similar to equation (5.6). Then the term corresponding to case (b) is obtained by considering the two following components:

(a) a_j customers in node j, which is its minimum queue length, and the remaining $n - a_j$ customers in the first j-1 nodes,

(b) at least one customer in node j and the remaining n-1 customers in the first j nodes (node j included).

Case (a) and case (b) correspond respectively to the first and to the second term of the right hand side of equation (5.5).

Equation (5.6) has the same meaning of equation (5.5), but without the term corresponding to case (a). Indeed, the number of customers is $n > a_j + B(j-1)$ and such that node j cannot attain its minimum queue length a_j even if all the first j-1 nodes have the maximum queue length (i.e., with B(j-1) customers in the first j-1 nodes).

Equation (5.7) applies when the number of customers exceed the minimum overall capacity of the first j nodes (A_{j-1}) and node j capacity (B_j), i.e., for $n > A_{j-1} + B_j$. This equation is similar to equation (5.6). Then the term corresponding to case (b) is obtained by considering the two following components:

(b1) at least one customer in node j and the remaining n-1 customers in the first j nodes, given by the term $\rho_j G_j(n-1)$. Note that the distribution of these n-1 customers includes the states where $n_j = B_j$. Hence the overall distribution of the n customers includes unfeasible states where $n_j = B_j + 1$ and the remaining $n - B_j - 1$ customers are in the first j-1 nodes;

(b2) we discard these unfeasible states. This corresponds to the negative term

$$-\rho_j^{B_j+1} G_j(n - B_j - 1).$$

Equation (5.9) can be interpreted as equation (5.7) by noting that the states where node j shows its minimum queue length ($n_j = a_j$) are feasible, which correspond to case (a) above. Hence the contribution of these states to function $G_j(n)$ is given by the first term of the right hand side of equation (5.9).

The convolution algorithm for queueing networks with blocking is illustated in Table 5.3.

The convolution algorithm has a linear time computational complexity in the number of network components: it requires O(M N) operations, and more precisely O(MC), where $C = \max\{B_j - a_j \ 1 \le i \le M\}$.

As concerns numerical stability of the convolution algorithm, for networks with infinite capacity queues it is known that depending on the choice of the parameters, some values of function $G_j()$ may exceed the floating point range of some computers, or for constant near to the lower magnitude limit of the floating point range, one can observe truncation errors which lead to inaccuracy in the computation. Some static and dynamic scaling techniques can be used to substantially reduce this problem, although it cannot be completely eliminated. Similarly, for the convolution algorithms for networks with finite capacity queues possible numerical problems can be observed depending on the parameter values and can be alleviated by static and dynamic scaling techniques.

Table 5.3. Convolution Algorithm

[Computation of the values a_j, A_j, $B(j)$]
for j=1 to M
{

$$a_j = \max(0, N - \sum_{1 \le k \le M, \, k \ne j} B_k \;);$$

$$A_j = \sum_{1 \le i \le j} a_i \;; \quad B(j) = \sum_{1 \le i \le j} B_i$$

}
[initialisation]
for n =a_1 to B_1

$$G_1(n) = \rho_1^n \;;$$

[computation of the functions $G_j(n)$]
for j=2 to M
{
MIN = min $(A_{j-1}+B_j, B(j-1)+a_j)$;
MAX = max $(A_{j-1}+B_j, B(j-1)+a_j;)$;

$$G_j(A_j) = \rho_j^{a_j} \, G_{j-1}(A_j-1) \;;$$

for n= A_j +1 to min(MIN, N)

$$G_j(n) = \rho_j^{a_j} G_{j-1}(n - a_j) + \rho_j \, G_j(n - 1) \;;$$

for n= MIN +1 to min(MAX ,N)
    if MIN = $B(j-1)+a_j$

        then $G_j(n) = \rho_j \, G_j(n-1)$;

        else

$$G_j(n) = \rho_j^{a_j} G_{j-1}(n - a_j) - \rho_j^{B_j+1} G_{j-1}(n - B_j - 1) + \rho_j G_j(n - 1)$$

for n=MAX +1 to min(B(j)-1, N)

$$G_j(n) = -\rho_j^{B_j+1} G_{j-1}\Big(n - B_j - 1\Big) + \rho_j \, G_j(n-1) \;;$$

if N > B(j) then $G_j(B(j)) = \rho_j^{B_j} \, G_{j-1}(B(j-1))$

}

5.2.2 Performance indices

From the convolution algorithm we can derive closed-form expressions for the marginal queue length probability and some average performance indices.

Let $G_{M-\{j\}}(m)$ denote the normalizing constant for the network with all the nodes except node j and m customers, $1 \leq j \leq M$, $0 \leq m \leq N$.

Theorem 5.2

The marginal queue length probability density of each node j, $1 \leq j \leq M$, can be written as a function only of $G_M(n)$, as follows:

if $A_M - B_M - a_j > B(M) - B_j + a_j$

$$\text{prob}\{n_j = k\} = \frac{\rho_j^{k-a_j}}{G_M(N)} \left[G_M(N-k+a_j) - \rho_j G_M(N-k+a_j-1) \right]$$

if $A_M - B_M - a_j < B(M) - B_j + a_j$

$$\text{prob}\left\{ n_j = k \right\} = \rho_j^k \, \frac{G_{M-\{j\}}(N-k)}{G_M(N)}$$

for $a_j \leq k \leq B_i$ and where $G_{M-\{j\}}(m)$ is given by the following Lemmas.

First we derive an intermediate result which states the relation between functions G_{M-1} and G_M. Then we apply this result to derive the relation between $G_{M-\{j\}}$ and G_M. For the proof and further details see the references reported in Section 5.5.

Let $d_M = B_M - a_M + 1$ and $p = \lfloor m - A_{M-1} - 2d_M + 1/d_M \rfloor$.

Lemma 5.1

i) if $A_{M-1} + B_M > B(M-1) + a_M$:

$$G_{M-1}(A_{M-1}) = \frac{1}{\rho_M^{a_M}} G_M(A_M) \tag{5.14}$$

and for $A_{M-1} + 1 \leq m \leq B(M-1)$

$$G_{M-1}(m) = \frac{1}{\rho_M^{a_M}} \left[G_M(m+a_M) - \rho_M G_M(m+a_M-1) \right] \tag{5.15}$$

ii) if $A_{M-1} + B_M \leq B(M-1) + a_M$

for $m = A_{M-1}$, $G_{M-1}(m)$ is given by formula (5.14);

for $A_{M-1} + 1 \leq m \leq A_{M-1} + B_M + a_M$, $G_{M-1}(m)$ is given by formula (5.15);

for $A_{M-1} + B_M - a_M + 1 \leq m \leq B(M-1) + a_M - B_M - 1$

     if $m > A_{M-1} + (p+1)d_M$

$$G_{M-1}(m) = \frac{1}{\rho_M^{a_M}} \sum_{i=0}^{p+1} \rho_M^{id_M} \left[G_M(m+a_M-id_M) - \rho_M G_M(m-1+a_M-id_M) \right]$$

if $m = A_{M-1} + (p+1)d_M$

$$G_{M-1}(m) =$$

$$\frac{1}{\rho_M^{a_M}} \sum_{i=0}^{p} \rho_M^{id_M} \left[G_M(m+a_M-id_M) - \rho_M G_M(m-1+a_M-id_M) \right] + \rho_M^{(p+1)d_M} G_M(A_M)$$

for $B(M-1) + a_M - B_M \le m \le B(M-1) - 1$

$$G_{M-1}(m) = \frac{1}{\rho_M^{B_M+1}} \left[\rho_M G_M(m+B_M) - G_M(m+B_M+1) \right]$$

$$G_{M-1}(B(M-1)) = \frac{1}{\rho_M^{B_M}} G_M(B(M))$$

Lemma 5.2
The following relationships between G_{M-1} and G_M in lemma 1 also hold between $G_{M-\{j\}}$ and G_M for any node j of the network, with the following substitutions:

ρ_M, a_M and B_M are respectively replaced by ρ_j, a_j and B_j
A_{M-1} is replaced by $A_M - a_j$
$B(M-1)$ is replaced by $B(M) - B_j$

Remark
The mean queue length of node j if $A_M - B_M - a_j > B(M) - B_j + a_j$ can be written as follows:

$$L_j = \frac{1}{G_M(N)} \sum_{k=a_j}^{B_j} k \rho_j^{k-a_j} [G_M(N-k+a_j) - \rho_j G_M(N-k+a_j-1)]$$

Note that if $A_M - B_M - a_j \le B(M) - B_j + a_j$, then the mean queue length of node j can be derived by considering the marginal queue length distribution given by theorem 5.2.

Other performance indices including node utilization, throughput and average response time are derived as defined in Section 2.5 of Chapter 2, by using product form expression (5.1) for the state probability.

Finally, another interesting performance index is the average busy period during which node j is full, denoted by b_j. This busy period is defined as the time from an arrival that makes node j full to the next departure from node j.

Let $v_j = \sum_{i=1, i \neq j}^{M} \mu_i p_{ij}$, $1 \leq j \leq M$.

Theorem 5.3

The busy period of node j, $1 \leq j \leq M$, can be written as a function of $G_M(n)$ as follows:

if $A_M - B_M - a_j > B(M) - B_j + a_j$

$$b_j = \frac{\Delta_j}{G_M(N - B_j + a_j + 1) - \rho_j G_M(N - B_j + a_j)} \frac{\rho_j}{v_j}$$

where

$$\Delta_j = G_M(A_M) \qquad\qquad\qquad\qquad \text{if } N - B_j = A_M - a_j$$

$$\Delta_j = G_M(N - B_j + a_j) + \rho_j G_M(N - B_j + a_j - 1) \qquad \text{if } N - B_j > A_M - a_j$$

if $A_M - B_M - a_j \leq B(M) - B_j + a_j$,
when the non-empty condition holds and $B(M) \geq N + 2$

$$b_j = \frac{\rho_j G_M(N) - G_M(N+1)}{\rho_j G_M(N+1) - G_M(N+2)} \frac{\rho_j}{v_j}$$

when the non-empty condition does not hold

$$b_j = \frac{G_{M-\{j\}}(N - B_j)}{G_{M-\{j\}}(N - B_j + 1)} \frac{\rho_j}{v_j}$$

where $G_{M-\{j\}}$ is given by lemma 2.

Note that in the last case $B(M) \geq N + 2$ because we also have to compute $G_M(n)$, for $n = N+1, N+2$.

Example 5.1. Consider a queueing network which satisfies the condition of product form in case I of formula (5.2). In particular, we assume that the network is a central server model with $M=8$ service centers and $N = 50$ customers and the following routing probabilities: $p_{1j}=1/7$, $2 \leq j \leq M$, $p_{i1}=1$, $2 \leq i \leq M$. The blocking mechanism is RS-RD and finite capacities $B_1=B_6=10$, $B_2=B_3=8$, $B_4=B_5=B_8=6$ and $B_7 =5$. Service rates are $\mu_1=5$, $\mu_2= \mu_3=\mu_6=8$, $\mu_4=\mu_7=2$ and $\mu_5=\mu_8=3$ customers per units of time.

The number of states of the underlying Markov model is 11210. By applying the proposed convolution algorithm, we compute the normalizing constant $G_8[50]=0.0126$ with $O(80)$ operations and we derive the average performance indices as shown in Table 5.4.

Example 5.2. Consider a cyclic queueing network with BBS-SO blocking which satisfies the condition of product form in case II of formula (5.2) with $M=10$ and $N = 78$ customers. Thus the network satisfies the non-empty condition (definition 5.1). Finite capacity are $B_1=B_6=B_{10}=8$, $B_2=10$, $B_3=B_4=B_9=7$, $B_5=B_7=B_8=9$ and service rates are $\mu_1=\mu_2=4$, $\mu_3=\mu_9=5$, $\mu_4=\mu_5=\mu_7=\mu_8=3$ and $\mu_6=\mu_{10}=3$ customers per units of times.

The number of states of the network Markov process is 715. The convolution algorithm requires O(40) operations and it computes the normalising constant $G_{10}[78] = 3.15\ e^{32}$, and the average performance indices shown in Table 5.5.

Table 5.4. Example 5.1: utilization, throughput, mean queue length and mean response time for each node

i	U_i	X_i	L_i	T_i
1	0.5401	2.7006	9.734	3.604
2	0.2132	0.2132	7.587	35.585
3	0.2132	0.2132	7.587	35.5585
4	0.2118	0.4237	4.780	11.279
5	0.2008	0.6025	3.342	5.547
6	0.2132	0.2132	9.587	44.966
7	0.2097	0.4194	3.834	9.141
8	0.2033	0.6099	3.546	5.814

Table 5.5. Example 2: utilization, throughput, mean queue length and mean response time for each node

i	U_i	X_i	L_i	T_i
1	0.2315	0.9261	7.304	7.886
2	0.2315	0.9261	9.717	10.492
3	0.1852	0.9261	6.717	7.253
4	0.3087	0.9261	6.783	7.324
5	0.3087	0.9261	8.597	9.282
6	0.4630	0.9261	7.597	8.202
7	0.3087	0.9261	8.304	8.966
8	0.3087	0.9261	8.597	9.282
9	0.1852	0.9261	6.597	7.123
10	0.4630	0.9261	7.783	8.403

5.3   SYMMETRICAL NETWORKS

In this Section, we consider a particular class of networks with blocking, the class of closed *symmetrical* exponential networks with finite capacities and blocking. In a symmetrical network each node is not distinguishable from the others. In particular, each service center has identical service time distribution and buffer capacity. Moreover, each service center has the same number of sending and destination nodes. Topologies such as ring, double ring, chordal ring and hypercube belong to this class and have been proposed as interconnection structures for multicomputer systems and for local area networks for reasons of design simplicity and load balancing.

We consider M-node closed queueing networks without loops, with a single type of jobs and N customers. Each node i, $1 \leq i \leq M$, has single server and finite buffer capacity, denoted by B_i. The service time at each node i, $1 \leq i \leq M$, is exponential, load independent with parameter μ_i and jobs are served according to an abstract service discipline (e.g. FIFO). Upon completion of a service at node i, a job goes to node j with probability p_{ij}. We assume $p_{ii}=0 \ \forall$ i. According to definitions given in Chapters 2 and 4, we denote by $P=[p_{ij}]$ the MxM routing matrix and by $S=(S_1,...,S_M)$ the system state at random time, where S_i represents the node i state ($1 \leq i \leq N$), whose definition depends on the type of blocking. State S_i always includes the number of customers in node i, n_i, $1 \leq i \leq M$.

We restrict our attention to the particular class of closed queueing networks, referred to as *symmetrical networks*, where each node is not distinguishable from the others. In particular, this means that:

(a) $\mu_i = \mu, B_i = B$ i=1, 2, ...M;

(b) $\forall$ i, $p_{1,i} \neq 0 \Rightarrow p_{1+m,((i+m-1) \bmod M)+1} \neq 0$ m=1, 2, ...M–1;

(c) $p_{ij} \neq 0 \wedge p_{ik} \neq 0 \Rightarrow p_{ij}=p_{ik}=r$,   i,j,k=1, 2, ...M.

If K is the outdegree of each node, note that $r = 1/K$, where r is the above defined common value of the non-zero entries of the routing matrix **P**.

Condition (b) guarantees that rows 2,..., M of the routing matrix are circular permutations of the first row.

Fig. 5.4 shows some possible topologies of symmetrical networks, that include the cyclic and the fully connected topologies.

For this class of networks, we consider the blocking types RS-RD, BBS-SO, BBS-O and BAS defined in Chapter 4. Given the assumptions of symmetry and of exponential service time distributions, the node i state S_i can be simplified as follows:

$S_i = n_i$ for RS-RD, BBS-SO and BBS-O blocking

$S_i = (n_i, s_i)$, $s_i = 0, 1$,

where s_i indicates whether the server is active ($s_i=1$) or blocked ($s_i=0$).

This is because the nodes are indistinguishable and so there is no need to distinguish among destination nodes for BBS blocking, and among sending nodes to be unblocked for BAS blocking (see definitions in Chapter 4).

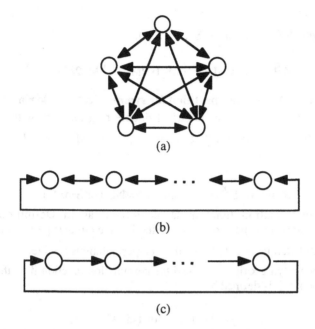

Figure 5.4. Examples of possible topologies for symmetrical networks: (a): complete connection; (b): double ring; (c): ring or cyclic topology

5.3.1 Solution reduction technique

In this Section we present a reduction solution technique that leads to the definition of a process with a reduced state space whose solution provides the steady state probabilities of the original system. This technique is based on a state space partition that groups the states with the same steady state probabilities.

Let us define the following state space partitions:

Definition 5.6. State space partition for any symmetrical network.
The state space E^X of a symmetrical network with X blocking ($X \in$ {RS-RD, BBS-SO, BBS-O, BAS}) can be partitioned in the subsets E_i^X such that:

$$S \text{ and } S' \in E_i^X \text{ if and only if } S = \Phi(S')$$

where the transformation function Φ simply performs a circular permutation of state S' and $|E_i^X| \leq M$ (the inequality holds if there are identical permutations). Function Φ can be defined by applying k times, $1 \leq k < M$ the permutation function α on the index set {1, ..., M}. Function α is defined by

$$\alpha(i) = (i+1) \bmod M \qquad (5.16)$$

For example, for a given state $(S_1, ..., S_M)$,

$$(S_1,..., S_M)=\Phi(S_M, S_1,..., S_{M-1})=\Phi(S_{M-1}, S_M,..., S_{M-2})=...=\Phi(S_2,..., S_M, S_1)$$

Definition 5.7. State space partition for symmetrical double ring networks.

The state space E^X of a symmetrical double ring network with X blocking ($X\in\{$RS-RD, BBS-SO, BBS-O, BAS$\}$) can be partitioned in the subsets E_i^X such that:

$$S \text{ and } S'\in E_i^X \text{ if and only if } S=\Phi(S') \text{ or } S=\vartheta(S')$$

where the transformation function Φ is defined as in Definition 5.1, the transformation function ϑ performs a so-called inverse circular permutation of state S' and $|E_i^X|\leq 2M$ (the inequality holds if some permutations are identical). Function ϑ is defined by applying k times, $1\leq k<M$ the permutation function β on the index set $\{1,...,M\}$. Function β is defined by

$$\beta(i)= \begin{cases} i-1 & \text{for } 1<i\leq M \\ \\ M & \text{for } i=1 \end{cases} \qquad (5.17)$$

Definition 5.8. State space partition for symmetrical complete connection networks.

The state space E^X of a symmetrical complete connection network with X blocking ($X\in\{$RS-RD, BBS-SO, BBS-O, BAS$\}$) can be partitioned in the subsets E_i^X such that:

$$S \text{ and } S'\in E_i^X \text{ if and only if } S=\Theta(S')$$

where the transformation function Θ performs a permutation of state S' and $|E_i^X|\leq M!$ (the inequality holds if some permutations are identical). Function Θ is defined by applying k times, $1\leq k<M$ the permutation function γ on the index set $\{1,..., M\}$. Function γ is defined by

$$\gamma(i)= j \text{ for any } j\in\{1, ..., M\} \qquad (5.18)$$

For example consider the network topologies in Figure 5.4 for M=4, N=5, and B=2. The state space E^X for $X\in\{$RS-RD, BBS-SO, BBS-O$\}$ is:

$E^X=\{(2,2,1,0), (2,2,0,1), (2,1,2,0), (2,1,1,1), (2,1,0,2), (2,0,2,1), (2,0,1,2), (1,2,2,0),$
$(1,2,1,1), (1,2,0,2), (1,1,2,1), (1,1,1,2), (1,0,2,2), (0,2,2,1), (0,2,1,2), (0,1,2,2)\}$.

If the network has a cyclic topology as in Figure 5.4-(c), the state space E^X may be partitioned according to Definition 5.6 as follows:

$$E^X = \{ E_1^X, E_2^X, E_3^X, E_4^X \}$$

where

$$E_1^X = \{(2,2,1,0), (0,2,2,1), (1,0,2,2), (2,1,0,2)\}$$

$$E_2^X = \{(2,2,0,1), (1,2,2,0), (0,1,2,2), (2,0,1,2)\}$$

$$E_3^X = \{(2,1,2,0), (0,2,1,2), (2,0,2,1), (1,2,0,2)\}$$

$$E_4^X = \{(2,1,1,1), (1,2,1,1), (1,1,2,1), (1,1,1,2)\}.$$

Note that $|E_i^X| = M \; \forall \; i$.

If the network has a double ring topology as in Figure 5.4-(b), the state space E^X may be partitioned according to Definition 5.7 as follows:

$$E^X = \{ E_1^X, E_2^X, E_3^X \}$$

where

$$E_1^X = \{(2,2,1,0), (0,2,2,1), (1,0,2,2), (2,1,0,2), (2,2,0,1), (1,2,2,0), (0,1,2,2), (2,0,1,2)\}$$

$$E_2^X = \{(2,1,2,0), (0,2,1,2), (2,0,2,1), (1,2,0,2)\}$$

$$E_3^X = \{(2,1,1,1), (1,2,1,1), (1,1,2,1), (1,1,1,2)\}.$$

Note that $|E_1^X| = 2M$ and $|E_i^X| < 2M$ for i=2, 3.

If the network has a complete connection topology as in Figure 5.4-(a), the state space E^X may be partitioned according to Definition 5.8 as follows:

$$E^X = \{ E_1^X, E_2^X \}$$

where

$$E_1^X = \{(2,2,1,0), (0,2,2,1), (1,0,2,2), (2,1,0,2), (2,2,0,1), (1,2,2,0), (0,1,2,2), (2,0,1,2),$$
$$(2,1,2,0), (0,2,1,2), (2,0,2,1), (1,2,0,2)\}$$

$$E_2^X = \{(2,1,1,1), (1,2,1,1), (1,1,2,1), (1,1,1,2)\}.$$

Note that $|E_i^X| < M! \ \forall \ i$.

In the following theorem we prove a property of the stationary joint queue length distribution of a symmetrical network. This property simply derives by the symmetry definition and it allows simplifying the process solution.

Theorem 5.4

For a symmetrical exponential network with state space E^X partitioned according to the above definitions in subsets E_i^X with stationary joint queue length distribution π, the following property holds:

$$\pi(S) = \pi(S') \qquad \text{if } S \text{ and } S' \in E_i^X$$

Proof.

For any symmetrical network the theorem is a simple consequence of the fact that the nodes are indistinguishable and the permutation functions (5.16), (5.17) and (5.18) guarantee that the population values of each pair of nodes sending-destination are the same for all states in E_i^X, disregarding the identity of the nodes. Note that the blocking phenomenon happens for particular values of population in a pair of sending-destination nodes. On the other hand, the stationary joint queue length distribution π depends on the state that is on the population of each sending-destination pair. However, because of the identity of the nodes, these values are the same for each state derived by the permutation functions (5.16), (5.17) and (5.18).

We sketch out the proof only for a double ring topology, since for the other topologies the proof is similar.

For example, let us consider a double ring network and a state $S=(n_1, ..., n_M) \in E_i^X$. According to the permutation function (5.16) the state $S'=(n'_1, n'_2, ..., n'_M)=(n_M, n_1, ..., n_{M-1})$ belongs to the same subset E_i^X.

Let us consider the sending-destination pair node1-node2 that, according to state S, has population values n_1 and n_2. In state S' one can find the sending-destination pair node2-node3 with population values $n'_2=n_1$ and $n'_3=n_2$. It is easy to be convinced that for each sending-destination pair with given population values according to S one can find another sending-destination pair that in S' has the same population values. Since the nodes are identical then states S and S' have the same stationary probabilities that is $\pi(S)=\pi(S')$. **QED**

By exploiting the result of Theorem 5.4, the stationary joint queue length distribution of a symmetrical network can be computed according to the algorithm illustrated in Table 5.6.

Figure 5.5 illustrates the reduction solution technique for a symmetrical double ring network with BBS-SO blocking.

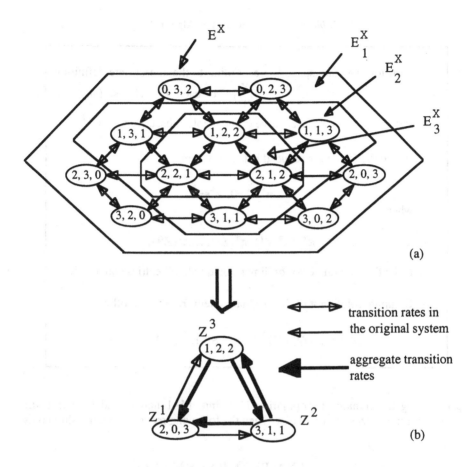

Figure 5.5. Reduction Solution Technique for a double ring symmetrical network with M=3, B=3, N=5 and X=BBS-SO: (a) original state diagram; (b) reduced state diagram

5.3.2 Performance indices

Given the symmetry assumption, the performance indices of different nodes are the same. In general, for all the considered blocking types, the performance indices are derived computing the state probabilities according to the reduction technique described in the previous section and utilizing standard numerical methods. Then the performance indices can be computed as defined in Section 2.6.

In the special case of a symmetrical and product form network with RS-RD blocking we can simplify the evaluation of some performance indices by using state enumeration as explained below.

To this end, note that the number of ways of partitioning N jobs into M nodes with capacity B, called I(N,M,B), may be obtained by setting x jobs at one node and

Table 5.6. Solution Reduction Algorithm

---

1. Partition the state space E^X according to the appropriate definition, that is $E^X = \{E_1^X, E_2^X, ..., E_q^X\}$. For each E_i^X choose (any) one state $S \in E_i^X$ as representative of the whole subset. Denote it Z^i.

2. Solve the following reduced balance system

$$\pi^* \hat{Q} = 0, \; \pi^* e = 1$$

   where

$$\pi^* = [\pi^*(Z^1), \pi^*(Z^2), ... \pi^*(Z^q)]$$

   and **e** the column vector of dimension q with all entries equal to 1.

3. Compute solution π of the original system from π^* as follows:

$$\pi(S) = \pi^*(Z^i) / |E_i^X| \qquad \forall \; S \in E_i^X, \; i=1, 2, ..., q.$$

---

partitioning the remaining population N−x into M−1 nodes, and repeating this operation for any feasible value of x. Then the following recursive relationship holds:

$$I(N,M,B) = \sum_{x=l}^{B'} I(N-x, M-1, B)$$

where $B' = \min\{N,B\}$, $l = \max\{0, N-(M-1)B\}$, $I(y,M,B) = 0$ if $y=0$, $I(N,1,B)=1$ and assuming that the value of the summation is zero when the lower extreme exceeds the upper extreme.

The following theorem states that the stationary state probability of a product form network with RS-RD blocking can be simply computed as the inverse of the state space cardinality.

Theorem 5.5

For a symmetrical network with RS-RD blocking and product form solution, the stationary state probabilities are identical for each state, that is:

$$\pi(S) = \pi(S') = \frac{1}{I(N,M,B)} \qquad \forall \; S, S' \in E^{RS-RD}$$

Proof

Under the hypotheses of the theorem, the product form solution is (see Section 5.1):

$$\pi(S) = \frac{1}{G} \prod_{i=1}^{M} \left(\frac{1}{\mu}\right)^{n_i} = \frac{1}{G}\left(\frac{1}{\mu}\right)^{\sum_{i=1}^{M} n_i} = \frac{1}{G}\left(\frac{1}{\mu}\right)^{N}$$

Hence $\pi(S) = \pi(S') \; \forall \; S, S' \in E^{RS\text{-}RD}$ and by the normalization condition

$$1 = \sum_{\forall S} \pi(S) = \left| E^{RS-RD} \right| \pi(S) = I(N,M,B)\pi(S)$$

from which the thesis follows. **QED**

By the symmetry assumption, the stationary queue length distribution can be defined as follows:

$$\pi_i(n_i) = \frac{\{\text{number of states with } n_i \text{ jobs at node } i\}}{\{\text{state space cardinality}\}} = \frac{I(N-n_i, M-1, B)}{I(N,M,B)}$$

As a consequence, the mean queue length may be simply computed by the usual expression:

$$L_i = \sum_{n_i=1}^{\min\{N,B\}} n_i \pi_i(n_i)$$

The other node performance indices that is utilization, throughput and mean response time are respectively defined as follows:

$$U_i^e = \text{Prob}\{\text{node } i \text{ is not empty and not blocked}\}$$

$$X_i^e = U_i^e \, \mu$$

$$T_i = L_i / X_i^e$$

In case of cyclic and double ring topologies (see Fig. 5.4 (c) and (b), respectively) with RS-RD blocking, the node utilization and the node throughput are symmetrical with respect to the maximum population that can be hosted in the network. This result is proved in the following theorem. Note that for cyclic topology the result holds also for networks without product form solution. Moreover, for this topology the result extends to blocking types {BBS-SO, BBS-O, RS-RD, RS-FD}, because in this case they are identical, as we shall see in Chapter 7.

Theorem 5.6

In a symmetrical homogeneous exponential network with cyclic topology and blocking type $X \in \{BBS\text{-}SO, BBS\text{-}O, RS\text{-}RD, RS\text{-}FD\}$ or double ring topology and RS-RD blocking type, the node utilization is symmetrical with respect to the maximum population that can be hosted in the network. In other words, denoting by $U_i(N)$ the node utilization when the network population is N, then

$$U_i^e(N) = U_i^e(MB-N) \qquad (5.19)$$

$$N=0, 1, ..., M\,B\text{-}1 \quad \text{for cyclic topology}$$
$$N=0, 1, ..., 2\,B\text{-}1 \quad \text{for double ring topology}$$

Proof

Consider a symmetrical exponential network with cyclic topology and blocking type $X \in \{BBS\text{-}SO, BBS\text{-}O, RS\text{-}RD, RS\text{-}FD\}$. Consider the dual network with MB−N holes as defined in Section 5.1. Note that partitioning N jobs into the M nodes is equivalent to partitioning MB−N holes into the same M nodes in the dual network. Hence I(N,M,B)=I(MB−N,M,B).

The self-duality property between the two networks can be proved. An interpretation of this property is that moving a job from one node to the next in the primal network means to move a hole from the next to the former node in the dual network. If in the primal network the destination node of node i is node i+1, in the dual network the destination node of node i is node i−1. Therefore the correspondence between jobs and holes yields the following state correspondence between the two networks:

$$(n_1, n_2, ..., n_M) \cong (B-n_1, B-n_2, ..., B-n_M)$$

where we use the symbol "$\cong$" to define state correspondence.

The steady state probability distribution of the two networks, denoted by π^P and π^d, are the same for corresponding states, that is:

$$\pi^P(n_1, n_2, ..., n_M) = \pi^d(B-n_1, B-n_2, ..., B-n_M)$$

By definition for the two networks we can write

$$U_i^e(N)=1- \sum_{\forall S \in E_{BL}} \pi^P(S), \qquad U_i^e(MB-N)=1- \sum_{\forall S \in E'_{BL}} \pi^d(S)$$

where E_{BL} and E'_{BL} are the subsets of the state spaces with the states where node i is empty or blocked. For example for node i=1 we have

$$E_{BL}=\{(n_1, n_2, ..., n_M) \in E(N)|\ (n_1=0) \lor (n_1 \neq 0 \land n_2=B)\}$$
$$E'_{BL}=\{(n'_1, n'_2, ..., n'_M) \in E'(MB\text{-}N)|\ (n'_1=0) \lor (n'_1 \neq 0 \land n'_M=B)\}$$

where $E^{RS\text{-}RD}(N)$ and $E'^{RS\text{-}RD}(MB\text{-}N)$ as the two space state for N jobs in the primal network and MB-N holes in the dual network.

Note that in the dual network the destination of node 1 is node M. Hence the statement (5.19) may be rewritten as

$$1- \sum_{\forall\, S\in E_{BL}} \pi^P(S) = 1 - \sum_{\forall\, S\in E'_{BL}} \pi^d(S)$$

that can be proved by proving that for each $S\in E_{BL}$ there exists a state $S'\in E'_{BL}$ with $\pi^P(S)=\pi^d(S')$. For example, consider the state $(0, n_2, ..., n_M)\in E_{BL}$. By using the state correspondence defined above and the results in literature (see Section 5.5), we have

$$\pi^P(0, n_2, ..., n_M)=\pi^d(B, n'_2, ..., n'_M)$$

By applying function Φ of Definition 5.6 and Theorem 5.4, we have also

$$\pi^d(B, n'_2, ..., n'_M)=\pi^d(n'_2, ..., n'_M, B)$$

Let us consider the state $(n'_2, ..., n'_M, B)$: if $n'_2=0$ then node 1 is empty, whereas if $n'_2\neq 0$ then node 1 is blocked. In both cases we have $(n'_2, ..., n'_M, B)\in E'_{BL}$ and

$$\pi^P(0, n_2, ..., n_M)=\pi^d(n'_2, ..., n'_M, B.)$$

For further details see the references mentioned in Section 5.5.

Consider a symmetrical exponential network with RS-RD blocking and double ring topology. Note that N<2B for the deadlock-free condition (see Section 2.2 of Chapter 2). Indeed if two nodes may be full simultaneously the deadlock can occur.

The node utilization is defined as follows:

$$U_i^e(N)=1- \sum_{\forall\, S\in E_{BL}} \pi^P(S)), \qquad U_i^e(MB-N)=1- \sum_{\forall\, S\in E'_{BL}} \pi^d(S)$$

where E_{BL} and E'_{BL} are the subsets of the state spaces with the states where node i is empty or blocked. For example for node i=1 we have

$$E_{BL}=\{(n_1, n_2, ..., n_M) \in E(N)|\ (n_1=0) \vee (n_1\neq 0 \wedge n_2=B) \vee (n_1\neq 0 \wedge n_M=B)\}$$
$$E'_{BL}=\{(n'_1, n'_2, ..., n'_M)\in E'(MB-N)\ |\ (n'_1=0) \vee (n'_1\neq 0 \wedge n'_2=B) \vee (n'_1\neq 0\wedge n'_M=B)\}.$$

For this network topology with RS-RD blocking the product form solution holds, as discussed in Section 5.1, and by Theorem 5.5 and $I(N,M,B)=I(MB-N,M,B)$, $\pi^P(S)=\pi^d(S')\ \forall\ S, S'$ with $S\in E(N)$, $S'\in E'(MB-N)$.

As a consequence, the statement (5.19) holds if E_{BL} and E'_{BL} have the same cardinality. Indeed by using the state correspondence defined above and the transformation functions of Definition 5.7, it is easy to define a bijective function between E_{BL} and E'_{BL}. **QED**

In Chapter 7 we will prove some monotonicity properties for a class of symmetrical networks.

5.4   ARRIVAL THEOREM

The arrival theorem for product form networks with infinite capacity provides the basic principle for the MVA computational algorithm. It states that the stationary state distribution at arrival instants of a customer at a particular node is equal to the stationary state distribution at arbitrary times of the same network, for open networks, and of the network with one less job, for closed networks. This result can also be applied for an efficient computation of the stationary state distribution at arrival times in the evaluation of passage time distribution, as discussed in Section 4.3 of the previous chapter.

An extension of the arrival theorem defined for networks with infinite capacity queues has been proved for a special class of networks in which a particular type of blocking can be defined by using the 'loss' and 'trigger' functions, which allow a constraint on the overall network population of a chain in multichain queueing networks with infinite capacity queues and is related to Recirculate blocking.

The proof of the arrival theorem for networks with infinite capacity queues is based on the BCMP product form solution. However, the arrival theorem cannot be directly extended to product form networks with blocking. Indeed, in networks with finite capacity queues the process associated to the network depends on the finite capacity of the queues, in particular both the process state space and the transition rate matrix. Then the behavior of a customer arriving at a node depends on the state of the network and in particular on the finite capacities, even if the network with blocking has a product form solution.

It is worth knowing that although in some cases the product form solution of the network with and without blocking are identical, up to a normalizing constant, the arrival theorem for networks with infinite capacity queues does not apply to the corresponding network with finite capacity queues.

For example, the direct application of the arrival theorem to a product form network with Stop protocol is shown to fail. Further results on the validity of the arrival theorem for product form queueing networks with some blocking types including Stop and Recirculate blocking were presented in literature (see references in Section 5.5).

However, the arrival theorem can still be extended to some cases of product form networks with blocking, but with a different interpretation.

Moreover some relations related to the arrival theorem hold even for non product form networks with BAS, BBS and RS blocking types queueing networks, based on the analysis of the arrival time distribution. Their application to the special case of some product form queueing networks provides an extension of the theorem with a different meaning than the networks without blocking.

These relations have been presented in Section 4.3 of the previous chapter. In particular Theorem 4.1 defines the stationary state probability distributions ξ at arrival instants at node i of a closed exponential network with finite capacity queues and blocking of type BAS, BBS-SO or RS-RD as a function of the stationary distribution at arbitrary times, π. Moreover, corollary 4.1 state an equality, up to a normalizing constant, between the arrival state probability and the probability at arbitrary time of the corresponding state (see Section 4.3 and the references in Section 5.5 for details).

From corollary 4.1 one can derive a relationship between the joint queue length distribution at arrival and arbitrary times of networks with different parameters for some closed exponential networks under BBS-SO, BAS and RS-RD blocking.

Consider closed networks with either a cyclic or central server topology. Let W denote the network model introduced above and let W^* denote a new network identical to W except for one less customer, and modified finite capacities denoted by B_j^*, $1 \leq j \leq M$. Let π^* denote the steady-state probability distribution at arbitrary times of network W*. Let A be the state space of the discrete time embedded Markov process at customer arrival time at node i. One can prove the following theorem.

Theorem 5.7
The stationary state distribution at arrival instants at node i of network W is identical to the state distribution at arbitrary times of network W*, i.e., $\forall S^e \in A$

$$\xi(S^e) = \pi^*(S^e)$$

i)   for product form networks with RS-RD or BBS-SO blocking and

   for a cyclic topology with $M \geq 2$ nodes and

   $B_j^* = B_j - 1$, for j=i,i-1 and $B_j^* = B_j$ for $j \neq i, i-1$, $1 \leq i, j \leq M$

   and for a central server topology with

   $B_j^* = B_j - 1$, for j=1,i and $B_j^* = B_j$ for $j \neq 1, i$, $1 \leq j \leq M$, $2 \leq i \leq M$

   where 1 denotes the central node;

ii) for the two-node product form network with BAS blocking and $B_j^* = B_j - 1$, for j=1,2.

This result has been applied to derive a closed form expression of the arrival time distribution and to compute passage and cycle time distribution in queueing networks with blocking, as described in Section 4.4 of Chapter 4.

This result on the arrival theorem can be extended to product form networks with Stop and Recirculate blocking.

Theorem 5.8
The stationary state distribution at arrival instants at node i of network W is identical to the state distribution at arbitrary times of network W*, i.e., $\forall S^e \in A$

$$\xi(S^e) = \pi^*(S^e)$$

i)   for product form networks with Stop blocking and for reversible routing topology with $M \geq 2$ nodes and

$$B_j^* = B_j - 1, \text{ for } j = i, i-1 \text{ and } B_j^* = B_j \text{ for } j \neq i, i-1, \ 1 \leq i, j \leq M$$

ii) for product form networks with Recirculate blocking and at most one full node and $B_j^* = B_j - 1$, for $1 \leq j \leq M$.

The extension of the arrival theorem to queueing network models with a more general topology and different blocking types, including heterogeneous networks where various node work under different blocking mechanisms, is an open issue.

5.5   BIBLIOGRAPHICAL NOTES

Product form Solution
The well-known class of product form BCMP queueing networks with infinite capacity queues is defined in Baskett et al. (1975). Several results have been presented in the literature for various special cases of product form networks with blocking. A survey of product form queueing networks with finite capacity queues and subnetwork population constraints, different blocking types and multiple classes of customers can be found in Balsamo and De Nitto Personè (1994).

Reversibility of queueing networks and of the underling stochastic process is discussed in Kelly (1979) as well as the Markov process truncation discussed in Section 5.1. In the same section we have introduced several definitions. The reader interested in details on definition 5.3 of condition (A) may refer to the particular model introduced by Akyildiz and Van Dijk (1990). It is a multiclass network with parallel queues with interdependent blocking functions and service rates, and which satisfy a so-called invariant condition. Product form formula PF6 is defined in the same work. Product form PF3 is given in Akyildiz and Von Brand (1989a).

Product form solution for both homogeneous and non-homogeneous two-node cyclic networks with blocking have been proved for exponential single class networks based on the reversibility property in Kingman (1969), Pittel (1979), Hordijk and Van Dijk (1981) and Akyildiz (1987). Product form solution for multiclass networks with BCMP nodes and blocking under additional constraints is derived in Onvural (1989), Choukri (1993) and Van Dijk and Tijms (1986). By using a similar approach, product form is derived for closed queueing networks with a reversible routing matrix and RS-RD or Stop blocking and different types of nodes in Kingman (1969), Pittel (1979) and Hordijk and Van Dijk (1981). Akyildiz and Von Brand (1989c), Onvural (1989), Yao and Buzacott (1985) and (1987) extend this result to include A-type nodes and more general blocking functions which may depend both on the total population, class population and routing chain population at the node. For central server networks with blocking product form is derived in Dallery and Yao (1986), Krzesinski (1987), Akyildiz and Von Brand (1989a), Towsley (1980) and Yao and Buzacott (1985).

State dependent routing and its relation with blocking networks is discussed in Krzesinski (1987) and Yao and Buzacott (1975) that extend to multiclass network the product form solution given in Towsley (1980). Akyildiz and Von Brand (1989c) generalize these results by combining state dependent routing and finite capacity queues.

Hordijk and Van Dijk (1983) introduce the job-local-balance property of the Markov process underlying the network model that yield product form solution. This balance property is related to routing reversibility and to local balance and station balance. The reader interested in such balance properties for queueing networks with infinite capacity are may refer to Baskett et al. (1975), Chandy and Martin (1983), Chandy, Howard and Towsley (1977), Cohen (1979), Kingman (1969) and Lam (1977).

Gordon and Newell (1967) introduce the concept of duality and derive product form solution PF2 of cyclic queueing networks with blocking by defining a dual network. This solution approach is extended to arbitrary topology networks with load independent service rates for RS-RD blocking in Hordijk and Van Dijk (1981) and to homogeneous networks with BBS-SO blocking under condition (B) and to heterogeneous networks in Balsamo and De Nitto Personè (1991). Duality is applied to exponential closed cyclic networks with RS-RD blocking in De Nitto Personè and Grillo (1987) and to cyclic networks with phase-type service distributions and BBS-SO blocking in Dallery and Towsley (1991) where they also derive some symmetry properties of performance indices.

Solution Algorithms

Product form BCMP queueing networks with infinite capacity queues can be analyzed by Convolution algorithm given in Buzen (1973) and Mean Value Analysis introduced in Reiser and Lavenberg (1980). A detailed definition for multiclass BCMP queueing networks can be found in Lavenberg (1983) and Kant (1992) and some static and dynamic scaling techniques to reduce the numerical instability of the Convolution algorithm in Lam (1977) and Lavenberg (1983).

The convolution algorithm described in Section 5.2.1 for a class of product form queueing networks with blocking is given in Balsamo and Clò (1998). A special case of MVA algorithm product form cyclic exponential network with BBS-SO and RS blocking is presented in Clò (1998). Another MVA algorithm for a class of product form queueing networks with RS blocking is given in Sereno (1999).

The algorithm proposed in Section 5.2 evaluates the average busy period. The definition and the analysis of the busy period of subnetworks in queueing models are introduced in Daduna (1988).

Symmetrical networks

Symmetrical queueing networks with blocking introduced in Section 5.3 with topologies such as ring, double ring, chordal ring and hypercube are used as models of interconnection structures for multicomputer systems and for local area networks as presented in Ree and Shwetman (1983) and Raghavendra and Silvester (1986).

The solution reduction algorithm for symmetrical networks presented in Section 5.3.1 are given in a preliminary form in De Nitto Personè and Grillo (1987). The proof of theorem 5.6 uses the self-duality property of the networks and other results defined in Gordon and Newell (1967).

Arrival Theorem

The arrival theorem for product form networks with infinite capacity in Lavenberg and Reiser (1980) and Sevcik and Mitrani (1981), as discussed in chapter 4. The arrival theorem is extended in Sevcik and Mitrani (1981) to networks with the special blocking defined by using the 'loss' and 'trigger' functions, similar to Recirculate blocking, that allow to represent constraints on the network population of a chain in multichain queueing networks with infinite capacity queues. A similar model is discussed in Van Dijk (1993).

The failure of the direct application of the arrival theorem to a product form network with Stop protocol is discussed in Van Dijk (1993). Boucherie and Van Dijk (1997) present further discussion on the validity of the arrival theorem for product form queueing networks with some blocking types including Stop and Recirculate blocking.

The arrival theorem for some special case of product form exponential networks with BBS-SO or BAS blocking is proved in Balsamo and Donatiello (1989a) and (1989b). Balsamo and Clò (1992) derive some relations related to the arrival theorem for non product form networks with BAS, BBS and RS blocking types queueing networks as presented in chapter 4, Section 4.3. They discuss the application of such relations to the special case of some product form queueing networks so obtaining an extension of the arrival theorem to networks with blocking with a different meaning than for networks without blocking.

The proof of theorem 5.7 is given in Balsamo and Donatiello (1989a), (1989b) and Balsamo and Clò (1992). Boucherie and Van Dijk (1997) extend this result to product form networks with Stop and Recirculate blocking.

REFERENCES

Akyildiz, I.F. "Exact product form solution for queueing networks with blocking" IEEE Trans. on Computer, Vol. 36 (1987) 122-125.

Akyildiz, I.F., and N. Van Dijk "Exact Solution for Networks of Parallel Queues with Finite Buffers" in *Performance '90* (P.J.B. King, I. Mitrani and R.J. Pooley Eds.) North-Holland, 1990, 35-49.

Akyildiz, I.F., and H. Von Brand "Central Server Models with Multiple Job Classes, State Dependent Routing, and Rejection Blocking" IEEE Trans. on Softw. Eng., Vol. 15 (1989) 1305-1312.

Akyildiz, I.F., and H. Von Brand "Computational Algorithms for Networks of Queues with Rejection Blocking" Acta Informatica, Vol. 26 (1989) 559-576.

Akyildiz, I.F., and H. Von Brand "Exact solutions for open, closed and mixed queueing networks with rejection blocking" J. Theor. Computer Science, Vol. 64 (1989) 203-219.

Balsamo, S., and C. Clò "State distribution at arrival times for closed queueing networks with blocking" Technical Report TR-35/92, Dept. of Comp. Sci., University of Pisa, 1992.

Balsamo, S., C. Clò "A Convolution Algorithm for Product Form Queueing Networks with Blocking" Annals of Operations Research, Vol. 79 (1998) 97-117.

Balsamo, S., and V. De Nitto Personè "Closed queueing networks with finite capacities: blocking types, product form solution and performance indices" Performance Evaluation, Vol. 12, 4 (1991) 85-102.

Balsamo, S., and V. De Nitto Personè "A survey of Product form Queueing Networks with Blocking and their Equivalences" Annals of Operations Research, Vol. 48 (1994) 31-61.

Balsamo, S., and L. Donatiello "On the Cycle Time Distribution in a Two-stage Queueing Network with Blocking" IEEE Transactions on Software Engineering, Vol. 13 (1989) 1206-1216.

Balsamo, S., and L. Donatiello "Two-stage Queueing Networks with Blocking: Cycle Time Distribution and Equivalence Properties", in *Modelling Techniques and Tools for Computer Performance Evaluation* (R. Puigjaner, D. Potier Eds.) Plenum Press, 1989.

Baskett, F., K.M. Chandy, R.R. Muntz, and G. Palacios "Open, closed, and mixed networks of queues with different classes of customers" J. of ACM, Vol. 22 (1975) 248-260.

Boucherie, R., and N. Van Dijk "On the arrival theorem for product form queueing networks with blocking" Performance Evaluation, Vol. 29 (1997) 155-176.

Buzen, J.P. "Computational Algorithms for Closed Queueing Networks with exponential servers" Comm. ACM, Vol. 16 (1973) 527-531.

Chandy, K.M., J.H. Howard, and D. Towsley "Product form and local balance in queueing networks"J. ACM, Vol.24 (1977) 250-263.

Chandy, K.M., and A.J. Martin "A characterization of product-form queueing networks" J. ACM, Vol.30 (1983) 286-299.

Choukri, T. "Exact Analysis of Multiple Job Classes and Different Types of Blocking" in *Queueing Networks with Finite Capacities* (R.O. Onvural and I.F. Akyidiz Eds.), Elsevier, 1993.

Cohen, J.W. "The multiple phase service network with generalized processor sharing" Acta Informatica, Vol.12 (1979) 245-284.

Clò, C. "MVA for Product-Form Cyclic Queueing Networks with RS Blocking" Annals of Operations Research, Vol. 79 (1998).

Daduna, H. "Busy Periods for Subnetworks in Stochastic Networks: Mean Value Analysis" J. ACM, Vol. 35 (1988) 668-674.

Dallery, Y., and D.D. Yao "Modelling a system of flexible manufacturing cells" in: Modeling and Design of Flexible Manufacturing Systems (Kusiak Ed.) North-Holland, 1986, 289-300.

Dallery, Y., and D.F. Towsley "Symmetry property of the throughput in closed tandem queueing networks with finite buffers" Op. Res. Letters, Vol. 10 (1991) 541-547.

De Nitto Personè, V., and D. Grillo "Managing Blocking in Finite Capacity Symmetrical Ring Networks" Third Int. Conf. on Data Comm. Systems and their Performance, Rio de Janeiro, Brazil, June 22-25 (1987) 225-240.

Gordon, W.J., and G.F. Newell "Cyclic queueing systems with restricted queues" Oper. Res., Vol. 15 (1967) 286-302.

Hordijk, A., and N. Van Dijk "Networks of queues with blocking", in: Performance '81 (K.J. Kylstra Ed.) North Holland (1981) 51-65.

Hordijk, A., and N. Van Dijk "Networks of queues; Part I: job-local-balance and the adjoint process; Part II: General routing and service characteristics", in: Lect. Notes in Control and Information Sciences (F. Baccelli and G. Fajolle Eds.) Springer-Verlag (1983) 158-205.

Kant., K. Introduction to Computer System Performance Evaluation. McGraw-Hill, 1992.

Kelly, F. P. Reversibility and Stochastic Networks. Wiley (1979).

Kingman, J.F.C. "Markovian population process" J. Appl. Prob., Vol. 6 (1969) 1-18.

Krzesinski, A.E. "Multiclass queueing networks with state-dependent routing" Performance Evaluation, Vol.7 (1987) 125-145.

Lam, S.S. "Queueing networks with capacity constraints" IBM J. Res. Develop., Vol. 21 (1977) 370-378.

Lavenberg, S.S. Computer Performance Modeling Handbook. Prentice Hall, 1983.

Lavenberg, S.S., and M. Reiser "Stationary State Probabilities at Arrival Instants for Closed Queueing Networks with multiple Types of Customers" J. Appl. Prob., Vol. 17 (1980) 1048-1061.

Onvural, R.O. "A Note on the Product Form Solutions of Multiclass Closed Queueing Networks with Blocking" Performance Evaluation, Vol.10 (1989) 247-253.

Pittel, B. "Closed exponential networks of queues with saturation: the Jackson-type stationary distribution and its asymptotic analysis" Math. Oper. Res. , Vol. 4 (1979) 367-378.

Raghavendra, C.S., and J.A. Silvester "A Survey of multi-connected loop topologies for local computer networks" Computer Networks and ISDN Systems, Vol. 11 (1986) 29-42.

Ree, D.A., and H.D. Shwetman "Cost-performance bounds for multicomputer networks" IEEE Trans. on Computer, Vol. 32 (1983) 83-95.

Reiser, M., and S.S. Lavenberg, "Mean Value Analysis of Closed Multichain Queueing Networks", J. ACM, Vol. 27 (1980) 313-322.

Sereno, M. "Mean Value Analysis of product form solution queueing networks with repetitive service blocking" Performance Evaluation, Vol. 36-37 (1999) 19-33.

Sevcik, K.S., and I. Mitrani "The Distribution of Queueing Network States at Input and Output Instants" J. of ACM, Vol. 28 (1981) 358-371.

Towsley, D.F. "Queueing network models with state-dependent routing" J. ACM, Vol. 27 (1980) 323-337.

Van Dijk, N. "On the Arrival Theorem for communication networks" Computer Networks and ISDN Systems, Vol. 25 (1993) 1135-1142.

Van Dijk, N., and H.G. Tijms "Insensitivity in two node blocking models with applications" in: Proc. Teletraffic Analysis and Computer Performance Evaluation, (Boxma, Cohen and Tijms Eds.) North Holland, 1986, 329-340.

Yao, D.D., and J.A. Buzacott "Modeling a class of state-dependent routing in flexible manufacturing systems" Annals of Oper. Res., Vol. 3 (1985) 153-167.

Yao, D.D., and J.A. Buzacott "Modeling a class of flexible manufacturing systems with reversible routing" Oper. Res., Vol. 35 (1987) 87-93.

6 APPROXIMATE AND BOUND ANALYSIS

In this chapter we deal with approximate and bound methods to analyze queueing networks with blocking and to evaluate various performance indices. Section 6.1 introduces the basic ideas of the approximate method proposed in the literature. Sections 6.2 and 6.3 present some approximate solution techniques for closed and open networks with blocking, respectively. Section 6.4 deals with bound approximation methods.

6.1 INTRODUCTION

As discussed in Chapter 4, exact analysis of queueing networks with blocking is based on the definition of an associated Markov process whose solution has an exponential space and time computational complexity in the number of system components, i.e., the number of service centers and the network population. Under special constraints for particular classes of networks we can define some efficient algorithms with polynomial time computational complexity. The two special cases of networks with blocking and product-form solution and symmetrical networks have been presented in Chapter 5.

We shall now consider networks with blocking that do not satisfy the constraints of these special classes, i.e. non product form and non symmetrical networks, and for which an exact solution based on the Markov process is prohibitively expensive.

In the literature many approximate techniques have been proposed to analyze open or closed queueing networks with finite capacity queues and population constraints in order to evaluate average performance indices and queue length distributions. Most of the methods provide an approximate solution with a limited computational cost, but without any bound on the introduced approximation error. The accuracy of the methods is usually validated by comparing numerical results with either simulation results or exact solutions. Few results have been presented for bounded aggregation, an approximate method that provides bound on the performance indices.

Most approximate methods are heuristics based on the *decomposition principle* applied to the underlying Markov process or directly to the network. Consider the discrete space continuous time homogeneous Markov process associated to a

queueing network with blocking introduced in Chapter 4, Section 4.1. Let E denote
the state space and Q the transition rate matrix. Decomposing a Markov process
consists in identifying a state space partition of E into K subsets E_k, $1 \le k \le K$, which
leads to a decomposition of the transition rate matrix Q into K^2 submatrices. Hence
the solution of the linear system of the global balance equations (4.4) to derive the
stationary state distribution π at arbitrary times is reduced to the solution of K
subsystems of smaller dimension, each related to a subset of E. Then these solutions
are combined to obtain the overall process solution.

More precisely, we can rewrite the stationary state probability $\pi(S)$ as the
product of the conditional probability of state S in E_k, $Prob(S \mid E_k)$, and the
aggregated probability of the subset E_k, $Prob(E_k)$, as follows:

$$\pi(S) = Prob(S \mid E_k) \, Prob(E_k) \qquad \forall S \in E_k, \, 1 \le k \le K$$

The decomposition technique substitutes the direct computation of $\pi(S)$ with the
computation of $Prob(S \mid E_k)$ and $Prob(E_k)$, for each S and E_k. Approximate
methods based on the decomposition of the Markov process provide an approximate
evaluation of the conditional and aggregated probabilities $Prob(S \mid E_k)$ and
$Prob(E_k)$.

A critical issue is the definition of the state space partition that affects both the
accuracy and the time computational complexity of the approximate algorithm.

The decomposition principle applied to the queueing network is based on the
aggregation theorem for queueing networks. It performs in three steps:

1. network decomposition into a set of subnetworks,
2. analysis of each subnetwork in isolation to define an aggregate component,
3. definition and analysis the new aggregated network obtained by substituting
   in the original network each subnetwork with the corresponding aggregated
   component.

Once the network partition is defined, the analysis of the original network
reduces to the analysis of each subnetwork at step 2 and of the aggregated one at
step 3. Note that network decomposition corresponds to a particular state space
partition of the underlying Markov process decomposition. Hence the subnetwork
analysis at step 2 corresponds to the evaluation of the conditional probability $Prob(S \mid E_k)$ and the analysis of the aggregated network at step 3 yields the evaluation of the
aggregated probability $Prob(E_k)$.

Network decomposition can be very efficient when the isolated subnetworks at
step 2 and the aggregated network are easy to analyze. For the special case of
product-form networks the aggregation theorem provides exact results, i.e., the
aggregated network is equivalent to the original model in terms of queue length
probability and average performance indices. In other words, the aggregated queue
length distribution and the average performance indices of the original network and
of the aggregated one are identical. This can be proved for product-form queueing
networks with infinite capacity queues and it holds also for queueing networks with
finite capacity queues and blocking. For non product-form networks the aggregated
network only approximates the original model, and no bound is known on the

approximation error. This result holds also for networks with infinite capacity queues and it is related to the general technique known as decomposition and aggregation method for Markov processes. Decomposability based on the Markov process can be applied to Markovian queueing networks and it can be related to subnetwork aggregation.

Many approximate methods for networks with blocking are based on the aggregation theorem and on network decomposition. Various heuristics have been defined by taking into account both the network model characteristics and the blocking type. A bounded aggregation technique defined for Markov processes can be applied to queueing networks with blocking by exploiting the special structure of the underlying Markov process.

The definition of the network decomposition and the evaluation of the approximation error are two critical issues for the approximate methods based on decomposition. Some techniques apply an iterative aggregation-disaggregation procedure for which conditions and speed of convergence should also be considered, as the second algorithm that we shall see for closed networks in the next section. Some approximations are obtained by forcing exact aggregation for product-form queueing networks with infinite capacity like the first algorithm for closed networks in the next section. Other approximation algorithms are based on a product-form solution defined by the maximum entropy principle that we describe in the next sections for closed and open networks, respectively.

Remark: Note that some algorithms defined for a particular model can be extended to a more general class of queueing networks with blocking by applying equivalence properties between models with different blocking types. They include heterogeneous models, i.e. networks where various service centers have different blocking mechanisms. Equivalence properties are presented in Chapter 7.

We distinguish two classes of approximate methods for closed and open networks. Indeed, one cannot immediately extend a method defined for one class of models to the other class. In the next two sections we will survey and compare some methods of the two classes. We classify the approximate methods by considering model assumptions, i.e. constraints on the network parameters such as topology, types of service distributions and blocking type. We assume single server nodes. We briefly recall each algorithm and discuss its rationale, the accuracy, the efficiency and the class of models to which it can be applied.

6.2 APPROXIMATE ANALYSIS OF CLOSED NETWORKS

In this section we survey some approximate methods for the performance analysis of closed queueing networks with finite capacity queues and blocking. We consider the following six algorithms based on various principles:

- Throughput Approximation
- Network Decomposition
- Variable Queue Capacity Decomposition

- Matching State Space
- Approximate MVA
- Maximum Entropy Algorithm

The first three methods analyze network with cyclic topology and the last three approximations apply to arbitrary topology networks. Table 6.1 shows the conditions under which the methods can be applied, and specifically the constraints on network topology, service centers (i.e. service time distribution, number of servers and queue capacity) and blocking type. Note that all the algorithms are applied to homogeneous networks, i.e. networks where each node has the same blocking type. We assume FCFS service discipline at each node. In the last column of Table 6.1 we report the key idea of each approximation method.

Let $B=\sum_{1\leq i\leq M} B_i$ denote the total network capacity and let $B^+=\max_{1\leq i\leq M}B_i$ and $B^-=\min_{1\leq i\leq M}B_i$ respectively denote the maximum and minimum node capacity of the network.

Table 6.1. Approximate methods for closed queueing networks with blocking

Method	Network Constraint		Blocking Type	Approximation key idea
	Topology	Node type		
Throughput Approximation	cyclic	G/M/1/B	BAS or BBS-SO	Exact model analysis for some network population. Interpolation of the throughput values by varying network population
Network Decomposition	cyclic	G/M/1/B	BBS-SO	Network decomposition into single nodes analyzed in isolation as M/M/1/B queues
Variable Queue Capacity Decomposition	cyclic a node with unlimited capacity	G/M/1/B	BBS-SO	Network aggregation of the set of finite capacity queue nodes in a single composite node having state dependent service rate and variable buffer size
Matching State Space	general	G/M/1/B	BAS	Analysis of the network with unlimited queue capacity and by choosing the network population to approximately match the same state space cardinality
Approximate MVA	general	G/M/1/B	BAS	Modification of the MVA algorithm to take into account blocking
Maximum Entropy Algorithm	general	G/GE/1/B	RS-RD	Approximate product-form for the queue length distribution

6.2.1 Approximate methods for cyclic networks

The first three methods, Throughput Approximation, Network Decomposition and Variable Queue Capacity Decomposition, evaluate the throughput of cyclic networks with exponential service time distribution. The first and the third algorithms compute the throughput as a function of network population.

Throughput Approximation

This method applies to cyclic networks with BBS or BAS blocking. We denote with BBS any of the BBS-SO, BBS-SNO and BBS-O blocking that yield the same behavior for cyclic networks, as discussed in Chapter 7.

This approximate method is based on the assumption that the throughput is a symmetrical function of the population network. Let $X(N)$ denote the network throughput when there are N customers in it. Throughput is symmetrical if $X(N)=X(B-N)$. This property holds for BBS blocking as proved under the more general assumption of phase-type service distributions, as discussed in Chapter 7. Moreover the throughput reaches its maximum value for $N=N^*$, where $N^*=\lfloor B/2 \rfloor$ if B is even and $N^*=\lfloor B/2 \rfloor, \lfloor B/2 \rfloor+1$ if B is odd. Function $X(N)$ is non-decreasing for $1 \leq N \leq N^*$ and non-increasing function for $N^* \leq N < B$. Hence the algorithm directly computes few values of function $X(N)$ with exact analytical methods and computes the other values by fitting the curve through those known points, by the following function

$$X(N)=X(N+1)-y\, x^{N^*-N} \tag{6.1}$$

where

$$y=[X(N^*) - X(B^-)]\,[\sum_{i=1}^{N^*-B^-} x^i]^{-1} \tag{6.2}$$

and x is the fixed-point of the following equation:

$$X(B^--1)= X(B^-)-[X(N^*)-X(B^-)][x^{N^*-B^-+1}(1-x)]/[x-x^{N^*-B^-+1}] \tag{6.3}$$

To overcome a throughput underestimation, a correction to formula (6.1) is proposed by adding a factor $c_N y$, where coefficient c_N is simply defined as follows (see the references given in Section 6.5 for further details):

$c_N = N^*-N$ if $N^*-1 \leq N \leq N^*--\lfloor (N^*-B^-)/2 \rfloor$

$c_N = c_{N-1}$ if $N=N^*-\lceil (N^*-B^-)/2 \rceil$ and $\lceil (N^*-B^-)/2 \rceil \neq \lfloor (N^*-B^-)/2 \rfloor$

$c_N = c_{N+1}-1$ if $B^-+1 \leq N \leq \lfloor (N^*-B^-)/2 \rfloor +1$

For BAS blocking the symmetry property of the throughput does not hold, but a similar shape of the curve as for BBS blocking is conjectured, supported by experimental results. However, N^* depends on the queue capacities and service rates and is approximated by $\lceil \Sigma_i (B_i+1)/2 \rceil -1$. In this case it is necessary to directly evaluate more values of X(N) than for BBS and to compute the approximation also for $N^*\leq N\leq B-2$. This affects the algorithm efficiency and accuracy with respect to the case of BBS blocking.

The approximate algorithm has the following structure:

1. Exact computation of X(N) for $N=B^--1$, B^-, N^*. Since the network for $1\leq N\leq B^-$ is without blocking, we can apply an algorithm for product-form networks without blocking. The exact evaluation of $X(N^*)$ requires the solution of the associated Markov chain, i.e. of linear system (4.4) (see Chapter 4, Section 4.2).
2. Approximate computation of X(N) for $B^-+1\leq N\leq N^*-1$. These values are approximated by formulas (6.1), (6.2) and the solution of the fixed-point problem (6.3).

For BAS blocking X(N) is not symmetrical and the algorithm has two additional steps:

3. Exact computation of X(N) for $N=B-1$, B. X(B) is evaluated as the average time between two successive deadlocks which are immediately detected and resolved; this requires the following numerical integration:

$$X(B) = \left(\int_0^\infty [1 - \prod_{i=1}^{M} (1-e^{-\mu_i t}) dt \right)^{-1}$$

X(B-1) is approximated by a function of X(B) or it can be directly computed. In the latter case note that the Markov process associated to the network has only $M*2^{M-1}$ states, because there are M choices for the unique non-full node and 2^{M-1} states for the remaining nodes, each being blocked or not.
4. Approximate computation of X(N) for $N^*+1\leq N\leq B-2$, as at step 2.

The algorithm is based on an iterative scheme for the fixed-point problem (6.3). Although convergence has not been proved, it has been observed. Space and time computational complexity of the algorithm mainly depends on the direct computation of X(N*) at step 1 obtained by the Markov chain analysis. As discussed in Chapter 4, the process state space cardinality grows exponentially with N and M and seriously affects the applicability of this method.

The algorithm shows a very good accuracy with a maximum relative error within 5% both for BAS and BBS blocking. For both blocking types the error increases when the number of customers is close to B^-. The width of the interpolation values does not affect the approximation accuracy. The main drawback of this method is the cumbersome computational complexity required to evaluate the exact throughput at steps 1 and 3. Hence, as observed by the authors, this method can be used for parametric analysis of the throughput by varying the network population and only for networks with a limited number of nodes and customers.

Network Decomposition

The throughput of the cyclic network with BBS blocking is approximated by a network decomposition method. In the first step the network is partitioned into M one-node subnetworks. At step 2 each subnetwork is analyzed in isolation as an $M/M/1/B_i$ network with arrival rate λ_i^* and load dependent service rate $\mu^*_i(n)$, $0 \leq n \leq B_i$ to derive the marginal queue length distribution $p^*_i(n)$, $0 \leq n \leq B_i$, $1 \leq i \leq M$. This aggregation procedure does not provide exact results for this blocking network and the analysis of the isolated queue is approximated by taking into account the blocking of customers due to the finite capacity of the downstream nodes. Let us consider two cases depending on whether all the nodes have finite capacity or there is one infinite capacity node, denoted by 1, i.e. $B_1 = \infty$. In the former case they propose the following definition of the parameters holds:

$$\mu^*_i(n) = \{(1/\mu_i) + \sum_{j=i+1}^{M} b_{ij}(n) [\sum_{k=i+1}^{j}(1/\mu_k)]\}^{-1} \quad 1 \leq i \leq M-1, \ 1 \leq n \leq B_i \quad (6.4)$$

$$\mu^*_M(n) = \mu_M, \ 1 \leq n \leq B_M \quad (6.5)$$

$$\lambda_i^* = X/(1-p^*_i(B_i)) \quad (6.6)$$

where X is the network throughput and $b_{ij}(n)$ denotes the probability that nodes $i+1, i+2, \ldots, j$ are full and node $j+1$ is not full, given n customers in node i, $1 \leq i,j \leq M$. This probability is expressed in terms of probabilities $p^*_k(B_k)$ for $i+1 \leq k \leq j+1$. Moreover $p^*_i(B_i)$ in formula (6.6) is a function of λ_i^*. Hence given the throughput X the arrival rate λ_i^* can be computed as the solution of the fixed-point equation (6.6).

These formulae are the basis of the iterative algorithm that starts with a throughput approximate interval $[X_{min}^{(0)}, X_{max}^{(0)}]$. At the k-th step $(k \geq 1)$ it computes new parameters λ_i^* and $\mu^*_i(n)$, $0 \leq n \leq B_i$, $1 \leq i \leq M$, by formulas (6.4)-(6.6) and appropriately updates the k-th throughput approximation $[X_{min}^{(k)}, X_{max}^{(k)}]$. The iterative scheme continues until a convergence condition is satisfied. These conditions include a control of the approximate interval width, i.e. $(X_{max}^{(k)} - X_{min}^{(k)}) < \varepsilon$ for a small ε, a consistency control, that is the summation of all the average node population is close to N, and $\lambda_i^* < \mu^*_{i-1}, \forall i$ for the convergence of the fixed-point equation (6.6). If all nodes have finite capacity $(B_1 < \infty)$ then also node M may be blocked by node 1. Hence formula (6.5) does not hold and $\mu^*_M(n)$ is computed as follows:

$$\mu^*_i(n) = \{(1/\mu_i) + \sum_{j=i+1}^{i-1} b_{ij}(n) [\sum_{k=i+1}^{j}(1/\mu_k)]\}^{-1} \quad 1 \leq i \leq M, \ 1 \leq n \leq B_i$$

$$(6.4')$$

where if i=M then i+1=1, if i=1 then i-1=M and we assume that

$$\text{if } j<i \text{ then } \sum_{k=i}^{j} z_k = \sum_{k=i}^{M} z_k + \sum_{k=1}^{j} z_k \; .$$

and an additional iteration cycle is required to compute probabilities $p^*_i(B_i)$, $\forall i$ (see references given in Section 6.5 for details).

Like the previous algorithm, convergence has not been proved, but it has been observed. One can show that the time computational complexity is of $O(kM^4(B^+)^3)$ operations for k iteration steps.

The algorithm has a good accuracy with a maximum relative error almost always within 7%. The the average relative error is affected by the number of nodes, but it seems to be independent of the other network parameters (service rate and capacity unbalancing). Like the previous algorithm, the error increases when the number of customers is close to B⁻.

Variable Queue Capacity Decomposition

This method can be applied to cyclic networks with BBS blocking and where one node has infinite capacity ($B_1=\infty$). The algorithm is based on the network decomposition principle applied to nested subnetworks. The key idea is that given a node i, all the downstream nodes $\{i+1,...,M\}$ are aggregated in a single composite node C_{i+1} with load dependent service rate and a *variable queue capacity*, as illustrated in Figure 6.1. This peculiar definition of variable queue capacity allows to overcome the classical definition of composite node with constant capacity, so improving the approximation. The approximation evaluates the composite node C_{i+1} parameters, that are the load dependent service rate denoted by $v_{i+1}(n)$ and the fraction of time in which the queue capacity is n, given N customers in the network denoted by $f_{i+1}(n|N)$, $1 \le n \le N$.

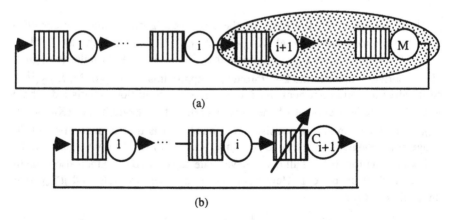

(a)

(b)

Figure 6.1. Aggregation of subnetwork $\{i+1,...,M\}$ in the composite node C_{i+1}: (a) the original network, (b) the aggregated network where node C_{i+1} has variable queue capacity

The algorithm starts with the analysis of the two-node subnetwork formed by $\{M-1,M\}$ to define the composite aggregate node C_{M-1}, that is seen by node M-2. Then the algorithm goes backward from node i=M-2 to node 1 to the analysis of the two-node subnetwork formed by $\{i, C_{i+1}\}$ to define the composite aggregate node C_i. At the last step the two-node network formed by $\{1, C_2\}$ represents the entire aggregated network and one obtains the approximated throughput.

The analysis of each two-node network with a composite node with variable queue capacity (or variable buffer) (VB) is carried out by considering two corresponding two-node networks with a composite node with fixed buffer (FB) and with infinite buffer (IB), respectively.

Let VB_i be the two-node network with nodes $\{i, C_{i+1}\}$ where the composite node C_{i+1} has variable buffer size and load dependent service rate. In particular the algorithm derives parameters $v_{i+1}(n)$ and $f_{i+1}(n|N)$ of node C_{i+1} in the VB_i network. These are obtained by the solution of two-node network VB_{i+1} formed by nodes $\{i+1, C_{i+2}\}$ and analyzed at the previous step.

In order to analyze each network VB_i we consider the two associated networks FB_i and IB_i introduced above. Let $X^{VB_i}(K)$, $X^{FB_i}(K|b)$ and $X^{IB_i}(K)$ denote respectively the throughput when there are K customers in each of the three networks and network FB_i has finite buffer size b. Similarly, for simplicity let C_{i+1} denote the composite node in all the three subnetwork, even if it has variable buffer size and variable service rate in VB_i, it has constant finite buffer size and variable service rate in FB_i and it has infinite buffer size and variable service rate in IB_j. Let $p_{C_{i+1},n}^{VB_i}(K)$, $p_{C_{i+1},n}^{FB_i}(K|b)$ and $p_{C_{i+1},n}^{IB_i}(K)$ denote respectively the marginal probability of n customers in node C_{i+1} when there are K customers in each of the three subnetworks, and network FB_i has finite buffer size b.

Then given the solution of network VB_{i+1}, that is given the throughput $X^{VB_{i+1}}(K)$ and probabilities $p_{C_{i+2},n}^{VB_{i+1}}(K)$ for $1 \le n \le K$ and each population $1 \le K \le N$, we can derive the parameters $v_{i+1}(n)$ and $f_{i+1}(n|N)$ of node C_{i+1} in the VB_i network as follows:

$$v_{i+1}(K) = X^{VB_{i+1}}(K) \qquad 1 \le K \le N \qquad (6.7)$$

$$f_{i+1}(n|K) = p_{C_{i+2},n-B_{i+1}}^{VB_{i+1}}(K) \qquad 1 \le n \le K-1, \ 1 \le K \le N \qquad (6.8)$$

$$f_{i+1}(n|K) = \sum_{j=0}^{B_{i+1}} p_{C_{i+2},n-B_{i+1}+j}^{VB_{i+1}}(K) \quad n=K, \ 1 \le K \le N \qquad (6.9)$$

where $p_{C_{i+2},n}^{VB_{i+1}}(K) = 0$ if n<0.

This is the basic step of the algorithm applied for i=M-1 to 1. Note that for i=M-1 we simply have $C_{i+1} = C_M$ and then the aggregated parameters are defined as follows:

$$v_M(K) = \mu_M \qquad (6.10)$$
$$f_M(n|K) = 1 \text{ if } n = B_M, \ 1 \le K \le N, \ f_M(n|K) = 0 \text{ otherwise}$$

The solution of each network VB_i is derived by the solution of the corresponding network FB_i that in turn is solved by the analysis of the IB_i networks. This last network has infinite capacity queues and its solution is well-known and quite simple. Then given the solution of network IB_i, that is given the throughput $X^{IB_i}(K)$ and probabilities $p^{IB_i}_{C_{i+1},n}(K)$ we define corresponding FB_i network. We assume that node C_{i+1} has finite buffer b and load dependent service rate $\mu^{FB_i}_{C_{i+1}}(n) = \mu^{IB_i}_{C_{i+1}}(n)$ if $n \le b$ and $\mu^{FB_i}_{C_{i+1}}(n) = \mu^{IB_i}_{C_{i+1}}(b+1)$ otherwise. Then the analysis of network FB_i leads to the following relations:

$$X^{FB_i}(K|b) = X^{IB_i}(K) \qquad \text{if } 1 \le K \le b \qquad (6.11)$$
$$X^{FB_i}(K|b) = X^{IB_i}(b+1) \qquad \text{if } K > b$$

$$p^{FB_i}_{C_{i+1},n}(K|b) = p^{IB_i}_{C_{i+1},n}(K) \qquad \forall n,\ 1 \le K \le b \qquad (6.12)$$
$$p^{FB_i}_{C_{i+1},n}(K|b) = p^{IB_i}_{C_{i+1},n}(b+1) \qquad 0 \le n < b,\ K > b$$
$$p^{FB_i}_{C_{i+1},n}(K|b) = p^{IB_i}_{C_{i+1},b}(b+1) + p^{IB_i}_{C_{i+1},b+1}(b+1) \qquad n = b,\ K > b$$
$$p^{FB_i}_{C_{i+1},n}(K|b) = 0 \qquad n > b,\ K > b$$

Hence from this solution we derive the performance indices of the network VB_i as a weighted sum of the FB indices as follows:

$$X^{VB_i}(K) = \sum_{k=1}^{K} X^{FB_i}(K|k)\, f(k|K) \qquad (6.13)$$

$$p^{VB_i}_{C_{i+1},n}(K) = \sum_{k=1}^{K} p^{FB_i}_{C_{i+1},n}(K|k)\, f(k|K) \quad \forall n \qquad (6.14)$$

The approximate algorithm has the following structure:

1. Initialization.
   Computation of parameters $v_M(K)$ and $f_M(n|K)$ of node C_M by formula (6.10).
2. For each node $i = M-1,\dots,1$
   For each population $K = 1,\dots,N$

   Solution of subnetwork VB_i
   2a For each population $j = 1,\dots,K$
   I. Solution of subnetwork IB_i with population j to obtain the throughput $X^{IB_i}(j)$ and probabilities $p^{IB_i}_{C_{i+1},n}(j)$, by applying the convolution algorithm for product form networks with infinite capacity queues.

II. Solution of subnetwork FB_i with population j to obtain the throughput $X^{FB_i}(j\,|K)$ and probabilities $p^{FB_i}_{C_{i+1},n}(j\,|K)$ by formulas (6.11) and (6.12).

2b Computation of the throughput $X^{VB_i}(K)$ and probabilities $p^{VB_i}_{C_{i+1},n}(K)$ of subnetwork VB_i by formulas (6.13) and (6.14).

2c If $i>1$ computation of parameters of node C_{i+1} in the VB_i network, that are $v_{i+1}(K)$ by formula (6.7) and $f_{i+1}(j|K)$ for $1 \leq j \leq K$ by formulas (6.8) and (6.9).

3. Provide the solution of the network throughput by $X^{VB_1}(K)$, for each population $K = 1, \ldots, N$.

The reader interested in further details of the algorithm may refer to the references given in Section 6.5. The algorithm is very simple, non-iterative and its time computational complexity is of $O(MN^3)$ operations.

Numerical evaluation shows accurate results for small networks (i.e. with $M=3,4$), but an increasing relative error as the number of nodes increases, although most of the relative errors are within 15%. By the algorithm definition one can expect that the accuracy worsen as the network dimension increases, because the final composite node C_1 represents $M-1$ aggregated nodes and each aggregation step introduces an approximation error. The number of customers do not affect the accuracy, but the algorithm underestimates the throughput for small populations and overestimates it for large ones. The average relative error seems to be independent of the unbalancing between node service rates and/or queue capacities, as observed in some specific test with stressed parameter values.

Comparison of Throughput Approximation, Network Decomposition and Variable Queue Capacity Decomposition

By comparing the numerical results of the three approximation algorithms we can derive some observations on accuracy, efficiency and generality:

- Network Decomposition (ND) is more accurate than Variable Queue Capacity Decomposition (VQCD) for both the average and the maximum relative error. This difference increases with the number of network nodes.
- Throughput Approximation (TA) is more accurate than ND for both the average and the maximum relative error. TA accuracy is more stable that ND as the number of network nodes increases.
- ND is more efficient than TA, which is limited to small networks.
- the time computational complexities of ND, given by $O(kM^4(B^+)^3)$, and of VQCD, $O(MN^3)$, show a different dependence on network parameters. We observe that if $N<MB^+$ then VQCD approximation is better than ND, while the opposite is true otherwise. This confirms that VQCD approximation is less efficient than the ND for large N.
- VQCD and TA provide the throughput for all the network population from 1 to N.

- ND is based on a fixed-point iteration and can show some numerical instability.
- ND and AT apply to a more general class than VQCD.

6.2.2 Approximate methods for arbitrary topology networks

The three methods Matching State Space, Approximate MVA and Maximum Entropy Algorithm apply to arbitrary topology networks. The first two methods assume BAS blocking and exponential service time and evaluate the network throughput. The third method assumes RS-RD blocking, generalized exponential service time and evaluates the queue length distribution and average performance indices.

Matching State Space

This method approximates the throughput of a network with BAS blocking and exponential service times. The basic idea is to approximate the behavior of the network with blocking with that of a network without blocking by choosing the population to approximately match the state space cardinality of the underlying Markov chain. The assumption is that the two networks with nearly the same state space cardinality should have similar throughputs.

Let M denote the Markov chain associated to the network with M nodes and blocking. Let K(N) denote the state space cardinality of M when there are N customers in the network. The algorithm defines a new network with infinite capacity queues and N' customers with underlying Markov chain M'. Let $K'(N')=\binom{M+N'-1}{M-1}$ be the state space cardinality of M' when there are N' customer in the network. The algorithm determines N' to approximate equation K(N)=K'(N'), that is to minimize the difference function $|K(N)-K'(N')|$. Since one can observe that K(N)≤K'(N) $\forall$N then N'≤N. Finally, the network without blocking is analysed. The algorithm has the following structure:

1. Computation of K(N) by a convolution algorithm.
2. Determine N' to minimize $|K(N)-K'(N')|$, 1≤N'≤N, by linear search in[1,N].
3. Computation of the throughput of the network without blocking by a convolution algorithm.

The algorithm implementation is simple and the time computational complexity is of $O(M^3+MN^2)$ operations. The approximation accuracy is fair with an average relative error of 4.6%, but a maximum relative error also more than 25% even for networks with few service centers. This can be explained by observing that the basic assumption of the approximation usually is not verified. The algorithm tries to approximate the state space cardinality independently of the model structure. Hence the algorithm provides the same approximation for all those networks with different parameters that have the same number of feasible states. The algorithm shows a good accuracy only for networks with very small state space, while the error increases with the state space cardinality. The methods provides more accurate results for central server networks and the worst case for cyclic networks. This could depend on a better approximation of the structure of the state space of the blocking network for the former topology than for the latter one. Indeed in a cyclic network we can observe a chain of up to M-1 blocked nodes and several simultaneous customer transitions, whereas in central server networks we can have at most two nodes involved in simultaneous transitions.

Approximate MVA

Network with BAS blocking and exponential service times are analyzed by a modification of the MVA algorithm originally defined for product-form networks with unlimited queue capacities. The MVA algorithm is based on the Little theorem and the arrival theorem. Note that the arrival theorem and the MVA algorithm as defined for networks without blocking cannot be immediately applied to networks with blocking, as discussed in chapters 4 and 5. Let $R_i(n)$, $L_i(n)$ and $X_i(n)$ denote respectively the average response time, mean queue length and throughput of node i when there are n customers in the network and let e_i be the mean number of visits at node i. For load independent service center MVA is based on the following recursive scheme:

$$R_i(n)= (1/\mu_i) [1+L_i(n-1)] \qquad 1\leq i\leq M \qquad (6.15)$$
$$X_i(n)= n\, e_i/[\Sigma_{1\leq j\leq M}\, e_j R_j(n)] \quad 1\leq i\leq M \qquad (6.16)$$
$$L_i(n)= X_i(n)R_i(n) \qquad 1\leq i\leq M \qquad (6.17)$$

for n=1,..., N, with $L_i(0)=0$, $1\leq i\leq M$. This algorithm is based on the arrival theorem defined for product-form networks that does not apply to networks with blocking. An exact modified MVA algorithm for a class of product-form networks with blocking has been defined in Chapter 5, Section 5.2.

The approximation algorithm modifies formula (6.15) trying to take into account blocking. In particular if node i is full it cannot accept new customers and there is at least one node j blocked by node i. The approximate algorithm modifies relation (6.15) for the full node i and for the blocked node j as follows:

$$R_i(n)= (1/\mu_i)\, L_i(n-1) \qquad\qquad (6.18)$$
$$R_j(n)= (1/\mu_j)\, [1+L_j(n-1)]+BT_i\, (e_j p_{ji}/e_i) \qquad (6.19)$$

where $BT_i= (1/\mu_i)$. For node i only the customers already in the node contribute to the average response time, while in the blocked node j the time increase of a blocking time due to node i. Then the modified MVA algorithm works as follows (see the references given in Section 6.5 for further details):

1. Initialization.
2. For each population n=1,...N
   <u>repeat</u>
   computation of the MVA equations by formulas (6.15)-(6.17), where (6.15) is substituted by (6.18) or (6.19) for full and blocked nodes, respectively
   <u>until</u> $L_i(n)\leq B_i$ for each node i.

The algorithm can be simply implemented and the time computational complexity is of $O(M^3+kMN)$ operations where k is the number of iterations of the internal cycle at step 2. Numerical experiments have shown small values of k. The algorithm does not show a fair accuracy for the average relative error that is over 10% for mean response time and about 8% for the throughput. The maximum relative error is more than 55% for mean response time and more than 38% for the throughput, even for networks with few service centers. Moreover for mean response time there is often a high variability in the relative errors of different service centers in the same network, while the approximation behavior is more regular for the throughput. The approximation error is not affected by the number of network nodes, but it depends on the network topology. Like the previous algorithm the algorithm provides better results for central server networks and the worst case for cyclic networks.

Comparison of Matching State Space and Approximate MVA

We can compare the throughput obtained by the two approximation algorithms Matching State Space and Approximate MVA since they apply to the same class of networks.

- Matching State Space (MSS) is more accurate than Approximate MVA (AMVA) both in terms of average and maximum relative errors.
- The approximations are quite different, since their rationales are not related.
- Both approximations seem to be independent of network parameters (number of nodes, service rates and queue capacities), but dependent on the topology. Specifically they provide better results for central server networks and worse results for cyclic networks.
- MSS is more efficient than AMVA.
- Both algorithms are stable.

Maximum Entropy Algorithm

This method evaluates the queue length distribution and average performance indices of a network with RS-RD blocking and generalized exponential service time. The approximation is based on the principle of maximum entropy and is an extension of the algorithm defined for open networks that we shall introduce in the next section. It has successively been extended to multiclass exponential networks.

Let $a_i=\max\{0, N-B+B_i\}$ denote the minimum number of customers in node i, $1 \le i \le M$. The algorithm approximates the joint queue length distribution $\pi(S)$ for each network state $S=(n_1,\ldots, n_M)$ by maximizing the entropy function

$$H(\pi)=-\Sigma_S \pi(S)\log(\pi(S))$$

subject to the following constraints:

(I)   (normalization) $\Sigma_S \pi(S)=1$

(II)   (u_i is the probability that node i has more than a_i customers)
$$\Sigma_{n_i>a_i}\pi_i(n_i)=u_i$$

(III)   (L_i is the mean queue length)

$$\Sigma_{a_i \le n_i \le B_i} \, h_i(n_i)\pi_i(n_i) = L_i$$

(IV)   (Φ_i is probability that node i is full)

$$\Sigma_{a_i \le n_i \le B_i} \, f_i(n_i)\pi_i(n_i) = \Phi_i$$

where $h_i(n_i) = \min\{0, n_i - a_i - 1\}$ and $f(n_i) = \max\{0, n_i - B_i + 1\}$.

By the Lagrange's method of undetermined multipliers the algorithm determines an approximation of $\pi(S)$ that has the following product form expression:

$$\pi(S) = (1/Z) \prod_{i=1}^{M} x_i(n_i) y_i^{h_i(n_i)} z_i^{f_i} \qquad (6.20)$$

where Z is a normalizing constant, $x_i(n_i) = 1$ if $n_i = a_i$ and $x_i(n_i) = x_i$ if $a_i < n_i \le B_i$, and x_i, y_i and z_i are the Lagrangian coefficients corresponding to constraints (II)-(IV). The network cannot be decomposed into single nodes and coefficients x_i, y_i and z_i do not have a closed form expression. The algorithm approximates the closed network with a *pseudo* open network without exogenous departures and arrivals. This open network is analysed by the approximation based on the same principle applied to open networks that we shall see in Section 6.3. However, that method considers the effective throughput of the nodes obtained by the traffic equations. Since the traffic equations for a closed network only define the relative throughput or arrival rate at each node, a unique solution has to be chosen arbitrarily. Then the algorithm for closed networks introduces the following additional constraint on the average queue lengths $N = \Sigma_i L_i$ and slight modifications to derive a solution for x_i, y_i and an approximation for z_i, $\forall i$. Then coefficients z_i are iteratively evaluated.

The algorithm details are given in the reference reported in Section 6.5.

The approximate algorithm has two phases, as follows:

1. Analysis of the corresponding pseudo open network with the approximate algorithm for open networks slightly modified to derive coefficients x_i, y_i and an approximation for z_i, $\forall i$.
2. Iterative evaluation of coefficients z_i by applying a convolution algorithm that computes network throughputs. In order to compute new solution at each iteration step we define a new value of coefficient z_i by the estimated node arrival rates λ_i, throughput X_i and average queue lengths L_i as follows:

$$z_i \leftarrow z_i \, (\lambda_i \, N / X_i) \, \Big[\sum_{j=1}^{M} \lambda_j L_j / X_j \Big]^{-1}$$

The iteration scheme is repeated until the following relation is satisfied:

$$\lambda_i / X_i = \text{constant for } 1 \le i \le M.$$

The time computational complexity of the algorithm for step 1 depends on the algorithm for open networks (see Section 6.3) and for step 2 is of $O(kM^2N^2)$ operations where k is the number of iterations. The algorithm shows a good accuracy for the average relative error within 4%. However, it can also produce results with relevant maximum relative errors up to 60% for throughput and mean queue length.

The topology and the symmetry of network parameters, that is the difference among the service rates and/or the finite capacities of the service centers do not affect the throughput accuracy. However, the throughput accuracy depends on the coefficient of variation (c.v.) of the service distributions. In particular the approximation error increases as the c.v.'s increase. This is due to the approximation introduced by the algorithm for open networks applied at step 2 that modifies the c.v. when it is different from 1, corresponding to exponential case. Moreover, the approximate results do not necessarily satisfy the traffic balance equations of the throughput and often find the exact solution when applied to product-form networks. Finally, there can be problems of convergence for cyclic networks with asymmetric parameters (service rates and queue capacities).

Note that this is the only method that provides an approximation of the queue length distribution.

Table 6.2 summarises some observations of the performance comparison of the six approximate methods for closed networks.

6.3   APPROXIMATE ANALYSIS OF OPEN NETWORKS

In this section we survey and compare some approximate methods for the performance analysis of open queueing networks with finite capacity queues and blocking. We consider the following four algorithms based on various principles:

- Tandem Exponential Network Decomposition
- Tandem Phase-Type Network Decomposition
- Acyclic Network Decomposition
- Maximum Entropy Algorithm for Open Networks

Table 6.3 shows the conditions under which the methods can be applied, i.e. the constraint on the network topology, the type of service centers (service time distribution, number of servers and queue capacity) and the blocking type. We assume FCFS service discipline at each node. The basic idea of each approximation method is given in the last column of Table 6.3.

6.3.1 Approximate methods for tandem networks

The first two methods, Tandem Exponential Network Decomposition and Tandem Phase-Type Network Decomposition, apply to tandem networks with exponential service time distribution, BAS blocking and evaluate the network throughput.

Tandem Exponential Network Decomposition

The throughput of the tandem network with BAS blocking is approximated by network decomposition method. The network is partitioned into M one-node subnetworks T(i), $1 \leq i \leq M$. Subnetwork T(i) represents the isolated node i and is

Table 6.2.
Performance comparison of approximate methods for closed queueing networks with blocking

Method	Performance Indices	Accuracy	Efficiency
Throughput Approximation	$X(N)$: network throughput as a function of the network population	Very good	Poor for networks with more than 5 nodes
Network Decomposition	X: network throughput	Good	Good
Variable Queue Capacity Decomposition	$X(N)$: network throughput as a function of the network population	Good for networks with up to 4 nodes, inaccurate otherwise.	Fair
Matching State Space	X_i: node throughput	Fair	Good
Approximate MVA	L_i: mean queue length X_i: node throughput T_i: node mean response time	Fair for throughput, poor for other performance indices	Very good
Maximum Entropy Algorithm	π_i: queue length distribution L_i: mean queue length X_i: node throughput R_i: node mean response time	Fair for all the performance indices.	Fair

Table 6.3. Approximate methods for open queueing networks with blocking

Method	Network Constraint		Blocking	Approximation
	Topology	Node type	Type	key idea
Tandem Exponential Network Decomposition	tandem	G/M/1/B	BAS	Network decomposition into single nodes analyzed in isolation as M/M/1/B queues
Tandem Phase-Type Network Decomposition	tandem	G/M/1/B	BAS	Network decomposition into single nodes analyzed in isolation as M/Cox/1/B queues
Acyclic Network Decomposition	acyclic	G/M/1/B	BAS	Network decomposition into single nodes analyzed in isolation as M/M/1/B queues
Maximum Entropy Algorithm	general	G/GE/1/B	RS-RD	Network decomposition into single nodes analyzed in isolation as M/M/1/B queues

analyzed as an M/M/1/B_i+1 network with arrival rate $\mu_u(i)$ and service rate $\mu_d(i)$ to derive marginal probability $\pi_i(n)$, $\forall n$, $1 \leq i \leq M$. The network decomposition is illustrated in Figure 6.2.

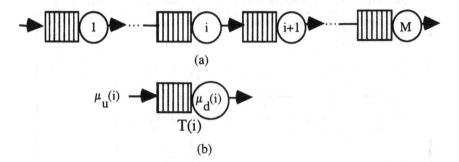

(a)

(b)

Figure 6.2. Network decomposition in M one-node subnetworks: (a) the original tandem network, (b) the isolated subnetwork T(i), $1 \leq i \leq M$

Since T(1) and T(M) correspond to the first and last node of the tandem network one has $\mu_u(1)=\lambda$ (exogenous arrival rate) and $\mu_d(M)=\mu_M$. The remaining 2(M-1) unknowns have to be determined. Let $p_b(i)$ denote the probability that at arrival time system T(i) is full and $p_s(i)$ the probability that at the end of a service system T(i) is empty. The approximation is based on the following relations:

$$\mu_u(i)=[(1/\mu_{i-1})+p_s(i-1)/\mu_u(i-1)]^{-1} \qquad 2 \leq i \leq M \qquad (6.21)$$

$$\mu_d(i)=[(1/\mu_i)+p_b(i+1)/\mu_d(i+1)]^{-1} \qquad 1 \leq i \leq M-1 \qquad (6.22)$$

$$X_1=X_2=...=X_M \qquad (6.23)$$

where equations (6.21) and (6.22) are obtained respectively by the analysis of the arrival and service processes at system T(i) and equations (6.23) by the throughput conservation law. This last set of equations can be rewritten in terms of the unknowns and the system state probability, because the throughput of system T(i) can be expressed as follows:

$$X_i = \mu_u(i)[1 - \pi_i(B_i+1)] = \mu_d(i)[1 - \pi_i(0)]$$

Each formula (6.21), (6.22) and (6.23) defines a set of M-1 equations and we can define three different systems of 2(M-1) to determine the unknowns $\mu_u(i)$ and $\mu_d(i)$, $1 \leq i \leq M$. The systems have been shown to be equivalent.

The algorithm has the following iterative scheme (see references in Section 6.5):

0.   Initialization: $\mu_u(1)=\lambda$, $\mu_d(i)=\mu_i$ $\forall i$.

1. <u>repeat</u>
    1a forward cycle: <u>for</u> i =1,..., M-1
        compute state probability π_i and probability $p_s(i)$ as follows:
          $p_s(i) = \pi_i(1)/(1 - \pi_i(0))$
        compute the service rate $\mu_u(i+1)$ by formula (6.21).
    1b backward cycle: <u>for</u> i=M,...,2
        compute probability $p_b(i)$ as follows:
          $p_b(i) = \pi_i(B_i)/(1 - \pi_i(B_i+1))$
        compute service rate $\mu_d(i-1)$ by formula (6.22).
    <u>until</u> $\max\{|X_i-X_j|, 1\leq i,j\leq M\}<\epsilon$. where ϵ is a predefined threshold.

2. Compute L_i, $\pi_i(n)$, $0\leq n\leq B_i$, $1\leq i\leq M$ and X_1.

The algorithm requires $O(kM(B^+)^2)$ operations where k is the number of iterations at step 1. One can prove the algorithm convergence, and numerical results show that it is fast. The accuracy is quite good for throughput with a maximum relative error within 4.1%. For the mean queue length the results are less accurate. The algorithm produces results with lower accuracy for parameters unbalancing, and in particular for unbalanced queue capacities.

Tandem Phase-Type Network Decomposition

Like the previous method this algorithm is based on network decomposition into M one-node subnetworks. The i-th subsystem corresponds to node i and is modelled by an $M/PH_n/1/B_i+1$ queue with arrival rate λ_i, phase-type service distribution with M-i+1 exponential phase to take into account blocking due to the downstream nodes $(i+1,...,M)$ and finite capacity B_i+1. Last subsystem is modelled by an $M/M/B_M+1$ queue with service rate μ_M. The i-th system $M/PH_n/1/B_i+1$ is solved with a matrix-geometric technique that compute a square matrix R_i. The phase-type distribution of system i can be represented by the pair (α_i,T_i), where $\alpha_i=[1,0,0,...0]$ is an (M-i+1)-vector, T_i is a square matrix defined as $T_i=[\mu_i, \mu_{i+1},...,\mu_M]^T A$, where $A[a_{rs}]$ $(i\leq r,s\leq M)$ is an upper triangular square matrix whose elements represent the probabilities among the exponential phases. The approximation evaluates these probabilities as follows:

$$a_{ij}=\varphi_{i+1}w_{i+1}(j) \qquad\qquad 1\leq i\leq M, i+1\leq j\leq M \qquad (6.24)$$

$$\varphi_i=\pi_i(B_i+1)/[\lambda_i w_i T_i^{-1}1] \qquad 2\leq i\leq M-1 \qquad (6.25)$$

$$w_i(j)=\pi_i(0)\alpha_i R_i^{B_i}\gamma_{ij} / \pi_i(B_i) \qquad 2\leq i\leq M-1, i\leq j\leq M \qquad (6.26)$$

where $w_i=[w_i(j)]$ $(i\leq j\leq M)$ is row (M-i+1)-vector, $1=[1...1]^T$ and γ_{ij} column (M-i+1)-vectors and γ_{ij} has all zero components except the j-th that is equal to one. Moreover the algorithm solves the fixed-point problem

$$\lambda_i = \lambda_i / [1 - \pi_i(B_i + 1)] \qquad\qquad 1 \le i \le M \qquad\qquad (6.27)$$

The algorithm for tandem networks where the first node has unlimited capacity $(B_1 = \infty)$ has the following scheme (see references in Section 6.5 for further details):

1. Analysis of subsystem M
   1a determine λ_M as the fixed point solution of equation (6.27)
   1b compute queue length distribution π_M by the $M/M/1/B_M + 1$ analysis
   1c compute $\varphi_{M-1} = \mu_M \pi_M (B_M + 1) / \lambda_M$
2. Analysis of subsystem i, i=M-1,..., 2
   2a determine λ_i as the fixed point solution of equation (6.27)
   2b compute queue length distribution π_i by the $M/PH_{M-i+1}/1/B_i + 1$ analysis
   2c compute $w_i(j)$, φ_i and $a_{i-1 j}$, $\forall j$, by formulas (6.26) (6.25) and (6.24)
3. Analysis of subsystem 1
   compute queue length distribution π_1 by $M/PH_M/1/B_1 + 1$ analysis with arrival rate λ.

The algorithm for tandem networks where all nodes have finite capacity $(B_1 < \infty)$ has an additional iterative cycle to estimate the effective arrival rates λ_i for each node i. In particular the first arrival rate is different from the exogenous arrival rate λ. At each iteration step the algorithm above for $(B_1 = \infty)$ provides a new value for the approximated arrival rates. Convergence has not been proved and there are cases of non convergence of the algorithm. This could depend on numerical errors due to the matrix dimension that grows with the number of nodes. The algorithm requires

$$O\left(k_1 \sum_{i=2}^{M} k_i (M-i+1)^3 B_i^2 \right)$$ operations where k_i is the number of iterations to

compute λ_i $\forall i$. These values depend on the fixed point problems at step 2 ($2 \le i \le M$) and on the external iteration cycle for i=1. Numerical results have shown a positive correlation between k_i, $\forall i$. The accuracy of the algorithm is good for the throughput and fair for the average queue length. The results are more accurate when all the nodes have finite capacity $(B_1 < \infty)$. The observed accuracy is affected by the parameter unbalancing (service rates and queue capacities), but does not depend on the number of nodes.

Comparison of the approximate methods for tandem networks
By comparing Tandem Exponential Network Decomposition (TEND) and Tandem Phase-Type Network Decomposition (TPND) defined for tandem networks with BAS exponential networks we can make some observations.
- The accuracy of the two methods is almost similar. TPND is slightly better than TEND for high blocking probabilities.
- The approximations obtained by TPND and TEND are quite similar, for sign and value.
- The approximation accuracy of TPND is influenced by capacity queue unbalancing, while TEND is affected by service rate unbalancing.

- Both TPND and TEND accuracy increases for small blocking probabilities, i.e. for networks with large queue capacities or large service rate with respect to the arrival rate.
- TEND is certainly more efficient than TPND.
- TPND can show numerical instability that can affect the algorithm convergence.
- The implementation of TEND is simpler than TPND.

6.3.2 Approximate methods for acyclic and arbitrary topology networks

The last two methods Acyclic Network Decomposition and Maximum Entropy Algorithm apply to more general topology networks, evaluate the queue length distribution and are respectively based on network decomposition and the maximum entropy principle. Acyclic Network Decomposition method applies to acyclic networks with exponential service time distribution and BAS blocking, while Maximum Entropy Algorithm analyses the more general classes of networks with arbitrary topology and generalized exponential service time distribution and RS-RD blocking.

Acyclic Network Decomposition

This method extends to acyclic topology networks the Tandem Exponential Network Decomposition defined for tandem networks and BAS blocking. Like that method this approach is still based on a network decomposition into M single node subsystems $T(i)$, $\forall i$, but defines a new set of equations to determine the subsystems parameters. Let λ_i be the exogenous arrival rate at node i and μ_i the service rate. Let I_i an indication function of such arrivals at node i defined as $I_i=0$ if $\lambda_i=0$, $I_i=1$ if $\lambda_i \neq 0$. Similarly let O_i be a function that denote whether customers can exit the network from node i, defined as $O_i=0$ if $p_{i0}=0$, $O_i=1$ if $p_{i0} \neq 0$. Then let $U_i=\{j: p_{ji}>0\} \cup I_i \{0\}$ and $D_i=\{j: p_{ij}>0\} \cup O_i \{0\}$ denote respectively the set of the indices of the predecessor and destination nodes of node i, $\forall i$, including node 0 for exogenous arrivals and departures.

Each subsystem $T(i)$ is illustrated in Fig. 6.3 and it is composed of the node i with finite capacity B_i and that receives arrivals from $|U_i|$ exponential sources with rates $\mu_{uj}(i)$, for $j \in U_i$. Note that these sources include exogenous arrivals at node i and arrivals from upstream nodes in the network. We assume that the ordering of the nodes is such that the routing probability matrix is upper triangular, that is $p_{ij}>0$ if and only if $i<j$, $\forall i,j$

$T(i)$ is analysed as an $M/M/1/B_i$ type network with arrival rate $\Sigma_{j \in U_i} \mu_{uj}(i)$ and service rate $\mu_d(i)$ to derive marginal probabilities $\pi_i(n)$, $\forall n$, $1 \leq i \leq M$. The state of system $T(i)$ is defined as the number of customers in node i plus the blocked

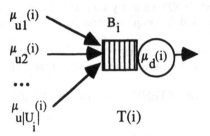

Figure 6.3. Network decomposition in M one-node subnetworks: the isolated subnetwork
T(i), $1 \leq i \leq M$

customers arriving from the $|U_i|$ streams. Thus B_i+k represents the state where k
upstream arrivals are blocked by node i, $0 \leq k \leq |U_i|$.

Let $\pi_i(n)$ be the state probability of system T(i), $0 \leq n \leq B_i+|U_i|$. This probability is
obtained by a birth-death Markov process with constant death rate $\mu_d(n)$ for each
state n and the following birth rate:

$$\mu^*(i)= \sum_{j=1}^{U_i} \mu_{uj}(i) \qquad \text{for } 0 \leq n \leq B_i-1$$

$$(k+1) \, \Omega_{k-1} / \Omega_k \qquad \text{for } n=B_i+k, \ 0 \leq k \leq |U_i|-1$$

where

$$\Omega_k = \sum_{j_1<j_2<...<j_k} \prod_{s=1}^{k} \mu_{u_{j_s}}(i), j_s \in U_i$$

In order to define the unknown rates $\mu_{uj}(i)$ and $\mu_d(i)$, $1 \leq i \leq M$, $j \in U_i$ we shall
now define some probabilities. Let $p_{bj}(i)$ be the probability that the j-th source is
blocked by node i, $j \in U_i$. Let $p_s(i)$ denote the probability that at the end of a service
system T(i) is empty that can be computed as follows:

$$p_s(i) = \pi_i(1) / [1-\pi_i(0)] \qquad (6.28)$$

The probability that at the end of a service at system T(i) the j-th source sees k
other sources blocked by node i, $j \in U_i$, $0 \leq k \leq |U_i|-1$ is denoted by $p_{bj}(k{:}i)$ and can be
computed as follows:

$$p_{bj}(k{:}i) = [\pi_i(B_i+k) - p_{bj}(k{:}i)] / [1 - p_{bj}(i)] \quad (6.29)$$

Then the approximation algorithm defines the unknown rates $\mu_{uj}(i)$ and $\mu_d(i)$,
$1 \leq i \leq M$, $j \in U_i$ as follows:

$$\mu_{uj}(i) = P_{ti}\Big[\frac{p_s(t)}{\mu^*(t)} + \frac{1}{\mu_d(t)} - P_{ti}\sum_{k=0}^{|U_i|-1} P_{bj}(k\!:\!i)\frac{k+1}{\mu_d(i)}\Big]^{-1} \qquad (6.30)$$

where t is the index of the j-th upstream node that is directly linked to node i.

If node i is the t-th upstream node which is directly linked to node m then we can write

$$\mu_d(i) = \Big[\frac{1}{\mu_i} + \sum_{m\in D_i} P_{im}\sum_{k=0}^{|U_m|-1} P_{bt}(k\!:\!m)\frac{k+1}{\mu_d(m)}\Big]^{-1} \qquad (6.31)$$

Formulas (6.30) and (6.31) define two sets of equations respectively obtained by the analysis of the arrival and service processes at system T(i). Moreover by the throughput conservation law we can write the following system of equations for throughput $X_{uj}(i)$ of the j-th source of node i and throughput $X_d(i)$ of system T(i)

$$X_{uj}(i) = X_d(k)\, p_{ki} \qquad (6.32)$$

where k is the index of the j-th upstream node of node i. This last set of equations can be rewritten in terms of the rates $\mu_{uj}(i)$ and $\mu_d(i)$, because the throughput of system T(i) can be expressed as follows:

$$X_{uj}(i) = \Big[\frac{1}{\mu_{uj}(i)} + \sum_{k=0}^{|U_j|-1}\frac{p_{bj}(k\!:\!i)(k+1)}{\mu_d(i)}\Big]^{-1}$$

$$X_d(i) = \Big[\frac{1}{\mu_d(i)} + \frac{p_s(k)}{\mu^*(i)}\Big]^{-1}$$

The three systems (6.30), (6.31) and (6.32) have been shown to be equivalent, like systems (6.21), (6.22) and (6.23) of the method for tandem networks.

The algorithm has the following structure:

0. Initialization
    $\mu_{u1}(i)=\lambda_i$ $\forall i$, such that $U_i = \varnothing$
    $\mu_d(i)=\mu_i$ $\forall i$, such that $D_i = \varnothing$
    $p_{bj}(n\!:\!i)=0$, $j\in U_i$, $0\leq n\leq|U_i|$, $\forall i$

1. repeat
    1a forward cycle: for i=1,..., M
        compute probability $p_s(i)$ by formula (6.28)
        compute rates $\mu_{uj}(i)$ $\forall j\in U_i$ by formula (6.30)
    1b backward cycle: for i=M,..., 1
        compute rate $\mu_d(i)$ by formula (6.31)
        compute probability $p_{bj}(k\!:\!i)$ $0\leq k\leq|U_i|-1$, $j\in U_i$, by formula (6.29)
    until convergence of $\mu_d(i)$, $\forall i$.

The algorithm time computational complexity is bounded by $O(kM[(U+B^+)^2+U^3+2^{U+1}])$ operations where k is the number of iterations and $U=\max_i|U_i|$. Note that for tandem networks U=1 and the algorithm requires $O(kM(B^+)^2)$ like the Tandem Exponential Network Decomposition algorithm. The algorithm shows a good accuracy with a maximum relative error within 5%. The accuracy of the algorithm can be analyzed by considering simple topologies like tandem, merge and split and more general acyclic topologies. Experimental results show that the accuracy is better for general acyclic topology and especially merge topology, while the worst results are for tandem networks. Hence the approximation works better when each subsystem has more than one predecessor subsystem. Another factor that affects the accuracy is the service rate unbalancing that worsens the results, while queue capacity unbalancing does not influence the accuracy. This is in contrast with the observed results for the Tandem Exponential Network Decomposition for tandem network, whose accuracy is affected by the queue unbalancing, and of which this method is a generalization.

Maximum Entropy Algorithm for Open Networks

Similarly to the method for closed networks described in the previous section, this technique approximates the queue length distribution of a network with RS-RD blocking and generalized exponential service time by using the principle of maximum entropy. For open networks the model is decomposed into M subsystems, each analysed as GE/GE/1/B nodes with appropriate parameters by considering blocking. In particular the approximate node i queue length distribution π_i maximizes the entropy function $H(\pi_i)=-\sum_n \pi_i(n)\log(\pi_i(n))$ subject to:

(I)   (normalization) $\sum_{0 \leq n \leq B_i} \pi_i(n)=1$

(II)   (utilization) $\sum_{0 \leq n \leq B_i} h_i(n)\pi_i(n)=\rho_i$

(III)   (L_i is the mean queue length) $\sum_{a_i \leq n \leq B_i} n\pi_i(n)=L_i$

(IV)   (Φ_i is probability that node i is full)

$$\sum_{a_i \leq n \leq B_i} f_i(n)\pi_i(n)=\Phi_i$$

where $h_i(n)=\min\{1, \max(0,n)\}$ and $f_i(n)=\max\{0, n-B_i+1\}$. This yields to the following product form expression for the joint queue length distribution:

$$\pi(n)=(1/Z) \prod_{i=1}^{M} x_i^{h_i(n_i)} y_i^{n_i} z_i^{f_i(n_i)}$$

where Z is a normalizing constant, $x_i=e^{-\beta_{i1}}$, $y_i=e^{-\beta_{i2}}$, $z_i=e^{-\beta_{i3}}$, and β_{ij} are the 3M Lagrange multipliers corresponding to constraints (II)-(IV). Finally the marginal queue length distribution for node i reduces to

$$\pi_i(n)=(1/Z') x_i^{h_i(n)} y_i^{n} z_i^{f_i(n)} \quad \forall n \qquad (6.33)$$

where Z' is a normalizing constant. The i-th system is analyzed as a GE/GE/1/B_i queue.

Let λ_i and c_i denote the arrival rate and the coefficient of variation of the interarrival time of all the customers arriving at node i, including those coming from other nodes and from node i itself. Let μ_i and c_{si} be the parameters of the service distribution of node i.

The algorithm computes various probabilities. Let π_{ij} be the probability that at service completion time at i node j is full and let π_{di} denote the probability that at service completion time node i is blocked.

Then the GE/GE/1/B_i queue has arrival parameters λ_i and c_i and service rate μ'_i and coefficient of variation c'_{si} defined as follows:

$$\mu'_i = \mu_i (1 - \pi_{di}) \qquad c'_{si} = \pi_{di} + c_{si} (1 - \pi_{di}) \qquad (6.34)$$

The arrival process at node i includes all the arrivals coming from outside, other nodes and node i itself because of blocking. The effective arrival process does not consider such arrivals from node i. Let λ'_i and c'_i denote respectively the service rate and the coefficient of variation of this process. Let π_i denote the probability that a customer arriving at node i sees the node full. Then the following relations hold:

$$\lambda_i = \lambda'_i / (1 - \pi_i) \qquad c'_i = (c_i - \pi_i) / (1 - \pi_i) \qquad (6.35)$$

The algorithm details are given in the references given in Section 6.5.
The approximate algorithm works as follows:

0. Initialization
   Computation of the initial modified service parameters μ'_i and c'_{si} defined as follows:
   $$\mu'_i = \mu_i (1 - p_{ii}) \qquad c'_{si} = p_{ii} + c_{si} (1 - p_{ii})$$
   Initialization of c_{di} and π_{0i}.

1. repeat
   for each subsystem i, $1 \leq i \leq M$
   1a compute the effective arrival rate λ'_i by the traffic equations with effective external arrival rate $\lambda'_{0i} = \lambda_{0i} (1 - \pi_{0i})$
   1b compute the probability π_{ij} that at service completion time at i node j is full, $\forall j$, by the non linear system defined by

$$\pi_{ij} = \frac{(1-s_{ji})^{B_i} r_i}{r_i(1-s_{ji})+s_{ji}} \pi_i(0) + \sum_{i=1}^{M} (1-s_{ji})^{B_i-n} \pi_i(n)$$

$$r_i = 2(1-c'_{si})$$

$$s_{ji} = 2[1 + (c_{dji} - \pi_{ji}) / (1 - \pi_{ji})]^{-1}$$

and with the additional relations:

$$\pi_i = [\lambda_{0i}\pi_{0i} + \sum_{j=1}^{M}(\lambda'_j p_{ji}\pi_{ji})/(1-\pi_{ji})] / [\lambda_{0i} + \sum_{j=1}^{M}(\lambda'_j p_{ji})/(1-\pi_{ji})]$$

$$c_{dj} = \rho'_j c'_{sj} + (1-\rho'_j)c'_j + (1-\rho'_j)\rho'_j, \quad \rho'_j = \lambda'_j/\mu'_j \qquad (6.36)$$

$$c_{dji} = 1 + p_{ji}(c_{dj} - 1), \quad c'_{0i} = \pi_{0i} + (1-\pi_{0i})c_{0i}$$

$$c'_i = -1 + \{\lambda'_{0i}/[\lambda'_i(c'_{0i}+1)] + \sum_{j=1}^{M}\lambda'_j p_{ji}/[\lambda'_i(c_{dji}+1)]\}^{-1}$$

1c   compute queue length probability π_i of the GE/GE/1/B_i queue by formula (6.33) where

$$x_i = \lambda_i \mu'_i \rho_i / [\lambda_i \rho_i (1-\mu'_i) + \mu'_i]$$
$$y_i = [\lambda_i \rho_i + \mu'_i (1-\lambda_i)]/[\lambda_i \rho_i (1-\mu'_i) + \mu'_i]$$
$$z_i = 1/[1-(1-\mu'_i)y_i]$$

and parameters λ_i, c_i, μ'_i and c'_{si} obtained by formulas (6.34) and (6.35)

1d   compute new values of c_{di} by formula (6.36)

<u>until</u>   convergence of c_{di}

The probability computation at step 1b requires the solution of the non-linear system that can lead to numerical instability and problems of convergence. There is no proof of convergence and uniqueness of the solution. The time computational complexity is of $O(\Omega^3)$, where Ω is the cardinality of the set of probabilities computed at step 1b. The algorithm's accuracy is fair. It decreases with the presence of cycles in the networks; merge and split topologies yield a good accuracy. The approximation error grows with the coefficient of variation. Indeed they are iteratively approximated to consider the influence of blocking and modified only if they are not one. Finally, note that the balance flow equations are not satisfied by the approximate throughput, because this constraint is not included in the maximum entropy problem.

Table 6.4 summarised some observations of the performance comparison of the six approximate methods for closed networks.

6.4   BOUND ANALYSIS

Bounded approximation determines an approximate solution with known accuracy or an upper and lower bound on the performance measure of interest. A few bounded approximation methods for Markovian queueing networks have been proposed. Some techniques are based on the analysis of the underlying Markov process and other methods consider particular properties of the network model.

The application of bounded approximation is twofold. An approximation value can be chosen in the interval obtained by the bounded method. This provides an approximate solution of the queueing network model with the advantage of known accuracy. Such approximation is reasonable when the bound interval width is small.

Table 6.4.
Performance comparison of approximate methods for open queueing networks with blocking

Method	Performance Indices	Accuracy	Efficiency
Tandem Exponential Network Decomposition	π_i: q. l. distribution L_i: mean queue length X_i: node throughput R_i: node mean resp. time	Very good for all performance indices	Very good.
Tandem Phase-Type Network Decomposition	π_i: q. l. distribution L_i: mean queue length X_i: node throughput R_i: node mean resp. time	Very good for all performance indices	Slow when applied to networks where all the nodes have finite capacity and fair otherwise.
Acyclic Network Decomposition	π_i: q. l. distribution L_i: mean queue length X_i: node throughput R_i: node mean resp. time	Very good for all performance indices	Very good.
Maximum Entropy Algorithm	π_i: q. l. distribution L_i: mean queue length X_i: node throughput R_i: node mean resp. time	Good for all performance indices	Fair.

Another application of bounded approximation is to provide an initial estimated value for other more complex approximation algorithms or define an interval to search for a solution with other approximate methods.

A bounded approximation is known as bounded aggregation for queueing networks and it is based on the decomposition principle applied to the underlying Markov process associated to the network. Let us consider a queueing network model and its underlying Markov process with state space E and transition rate matrix Q as defined in chapter 4 for various blocking types. Assume a network decomposition that defines a Markov process decomposition. Note that such decomposition depends on the performance measure of interest and it is a critical and non-trivial problem.

Bounded aggregation is defined for a Markov process whose state space is decomposed into K subsets E_k, $1 \le k \le K$, and the transition rate matrix Q into K^2 submatrices as described in section 6.1. The method derives lower and upper bounds on the stationary state conditional probabilities $Prob(\mathbf{S} \mid E_k)$ of a subset E_k, $\forall \mathbf{S} \in E_k$, from the submatrix of transition probabilities between the states of that subset only. The bounds can be improved when additional information on the entire Markov chain is available and for special structures of the submatrix, such as tridiagonal matrices, bounded approximation obtains analytical expressions of the bounds. The method can be applied to large Markov processes and derives bounds with various degrees of accuracy and computational complexity. However, Markov processes associated to queueing network models may not always yield tight bounds, although with additional complexity one can obtain bounds of higher orders.

An example of application of bounded aggregation is the evaluation of probability of customer rejection at a given node in an open queueing network with finite capacity queues and blocking. This model can be used for example to model communication protocols in packet-switched networks and in multicomputer interconnection networks. The performance measure of interest is the probability that an arrival at a given node i finds the buffer full and is rejected. This can be evaluated by computing the marginal probability that node i is full, that is $\pi_i(B_i)$. Then such probability can be derived from bounds on conditional probability by defining an appropriate state space partition of the underlying Markov process. State space E can be partitioned into subsets $E_k=\{S\in E: n_i=k\}$, $0\le k\le B_i$ and submatrices of the transition rate matrix can be easily defined by the definition given in chapter 4 for each blocking type. Then bounds on the rejection probability can be obtained by the aggregate probability of subset E_{B_i}, that is $\sum_{S\in E_k} Prob(S \mid E_k)$. The reader interested in further details on bounded aggregation may refer to the reference given in the next section.

Bounds on some average performance indices can be obtained for some classes of queueing networks with blocking. In particular bounds on the network throughput can be computed by using concavity and monotonicity properties, as we shall see in the next chapter.

6.5 BIBLIOGRAPHICAL NOTES

The aggregation theorem for single class queueing networks with infinite capacity queues and product form solution is given in Chandy, Herzog and Woo (1975). Related work can be found in Vantilborgh (1978). They consider aggregation of a subnetwork with single entry and exit points. Subnetwork aggregation for BCMP product form networks with infinite capacity queues is extended to multiclass networks in Kritzinger et al. (1982) and to arbitrary connected subnetworks in Balsamo and Iazeolla (1982).

The aggregation theorem for product form queueing networks with finite capacity queues and blocking and state-dependent routing is given in Balsamo and Iazeolla (1986). Boucherie and Van Dijk (1993) consider exact aggregation in product form networks with more general functions of state-dependent routing that may depend on the state of the entire network or of a subnetwork and/or single nodes. State dependent routing allows to represent systems with more complex features and in particular some blocking mechanisms, as discussed in chapter 1. Further extension of the aggregation theorem for queueing networks that include non product form components and state-dependent routing can be found in Boucherie (1998).

The decomposition principle and the introduction of the decomposition and aggregation method for Markov processes and its application to queueing network models is given in Courtois (1977). Decomposability applied to Markovian networks is discussed in Balsamo (1984) and Balsamo and Pandolfi (1984).

Several authors propose approximate methods for networks with blocking based on the aggregation theorem and on network decomposition by considering various network models and blocking types. Results are given in Hillier and Boling (1967), Gordon and Newell (1967), Konheim and Reiser (1978), Boxma and Konheim (1981), Yao and Buzacott (1985) and (1987), Suri and Diehl (1986), Perros and Altiok (1986), Altiok and Perros (1986) and (1987), Gershwin (1987), Brandwajin and Jow (1988), Perros et al. (1988), Akyildiz (1988a) and (1988b), Perros and Snyder (1989), Onvural and Perros (1989), Akyildiz and Perros (1989), Perros (1989), Dallery and Frein (1989) and (1993), Frein and Dallery (1989), Kouvatsos and Xenios (1989), Jun and Perros (1990), Lee and Pollock (1990), Mitra and Mitrani (1990) and (1992), Kouvatsos and Denazis (1993), Perros (1994), Kouvatsos and Awan (1995), Lee et al. (1995), Onvural (1990) and (1993), Cheng (1993), Bouchouch et al. (1996), Mishra and Fang (1997) and Balsamo and Rainero (1998).

Some approximate techniques apply an iterative aggregation-disaggregation procedure, like the algorithm for closed networks proposed in Dallery and Frein (1989) where conditions and speed of convergence are considered. Onvural and Perros (1989) present an approximation obtained by forcing exact aggregation for product-form queueing networks with infinite capacity. Approximation algorithms based on a product-form solution defined by the maximum entropy principle are given in Kouvatsos and Xenios (1989), Kouvatsos and Denazis (1993), Kouvatsos and Awan (1995).

Bounded aggregation for Markov processes applied to queueing networks with blocking is presented in Courtois and Semal (1986). Related results are given in Onvural and Perros (1989).

Bounds on the network throughput for closed cyclic networks and tandem open with BBS-SO and BAS blocking are given in Shantikumar and Yao (1989) and Liu and Buzacott (1993).

Approximation of Closed Networks with Blocking
The approximate method for cyclic closed queueing networks with blocking defined in Section 6.2.1 named Throughput Approximation is due to Onvural and Perros (1989), while Network Decomposition is given in Frein and Dallery (1989) and Variable Queue Capacity Decomposition is introduced in Suri and Diehl (1986). A generalization of Network Decomposition algorithm to general service time distribution where subsystems are analyzed with a two-moment approximation is given in Bouchouch et al. (1996). Related approximation methods for cyclic networks with BAS blocking can be found in Liu et al. (1989) and (1993). Approximate solution of tandem networks with Kanban blocking and generalized blocking are given in Mitra and Mitrani (1990) and (1992), Cheng (1993) and Mishra and Fang (1997).

The approximate algorithms named Matching State Space and Approximate MVA introduced in Section 6.2.2 for closed queueing networks with arbitrary topology networks are defined in Akyildiz (1988a) and (1988b).

Maximum Entropy Algorithm for closed networks with blocking is introduced in Kouvatsos and Xenios (1989) and its extension to multiclass exponential networks is given in Kouvatsos and Denazis (1993), Kouvatsos and Awan (1995).

Approximation of Open Networks with Blocking
The approximate method for tandem networks with blocking defined in Section 6.3.1 named Tandem Exponential Network Decomposition is given in Dallery and Frein (1993) and Tandem Phase-Type Network Decomposition is introduced in

Perros and Altiok (1986). Other approximate methods for tandem networks with BAS and BBS-SO blocking are given in Hillier and Boling (1967), Perros and Altiok (1986) and Brandwajin and Jow (1988).

Acyclic Network Decomposition to approximate acyclic networks introduced in Section 6.3.2 is defined in Lee, Bouchouch, Dallery and Frein (1995).

Finally, Maximum Entropy Algorithm for Open Networks is given in Kouvatsos and Xenios (1989).

REFERENCES

Akyildiz, I.F. "On the exact and approximate throughput analysis of closed queueing networks with blocking" IEEE Trans. on Soft. Eng., Vol. 14 (1988) 62-71.

Akyildiz, I.F. "Mean value analysis of blocking queueing networks, IEEE Trans. on Soft. Eng." Vol. 14 (1988) 418-129.

Akyildiz, I.F., and H.G. Perros, Special Issue on Queueing Networks with Finite Capacity Queues, Performance Evaluation, Vol. 10(1989).

Altiok, T., and H.G. Perros "Open networks of queues with blocking: split and merge configurations" IEE Trans. 9 (1986) 251-261.

Altiok, T., and H.G. Perros "Approximate analysis of arbitrary configurations of queueing networks with blocking" Ann. Oper. Res. 9 (1987) 481-509.

Balsamo, S. "Decomposability for General Markovian Networks", in *Mathematical Computer Performance and Reliability*, (G. Iazeolla, P.J. Courtois, A. Hordijk Eds.), North Holland, 1984.

Balsamo, S., and B. Pandolfi "Bounded Aggregation in Markovian Networks" in *Computer Performance and Reliability* (G. Iazeolla, P.J. Courtois, O. Boxma Eds.), North Holland, 1988.

Balsamo, S., and G. Iazeolla "An extension of Norton's Theorem for Queueing Networks" IEEE Transactions on Software Engineering, Vol.8 (1982) 298-305.

Balsamo, S., and G. Iazeolla "Synthesis of Queueing Networks with Block and State-dependent Routing", Computer Systems Science and Engineering, Vol.1 (1986) 194-199.

Balsamo, S., and A. Rainero "Approximate Performance Analysis of Queueing Networks with Blocking: A Comparison" UDMI/05/98/RR, Dept. Math and Comp. Sci., University of Udine, March 1998.

Boxma, O., and A.G. Konheim "Approximate analysis of exponential queueing systems with blocking" Acta Informatica, Vol. 15 (1981) 19-66.

Boucherie, R. "Norton's Equivalent for queueing networks comprised of quasireversible components linked by state-dependent routing" Performance Evaluation, Vol. 32 (1998) 83-99.

Boucherie, R., and N. Van Dijk "A generalization of Norton's theorem for queueing networks" Queueing Systems, Vol. 13 (1993) 251-289.

Bouchouch, A., Y. Frein and Y. Dallery "Performance evaluation of closed tandem queueing networks with finite buffers" Performance Evaluation, Vol. 26 (1996) 115-132.

Brandwajn, A., and Y.L. Jow "An approximation method for tandem queueing systems with blocking" Operations Research, Vol. 1 (1988) 73-83.

Chandy, K.M., U. Herzog and L. Woo "Parametric analysis of queueing networks" IBM Journal of Research and Development, Vol.1 (1975), 36- 42.

Cheng, D.W. "Analysis of a tandem queue with state dependent general blocking: a GSMP perspective" Performance Evaluation, Vol. 17 (1993) 169-173.

Courtois, P.J. *Decomposability: Queueing and Computer System Applications*, Academic Press, Inc, New York, 1977.

Courtois, P.J., and P. Semal "Computable bounds for conditional steady-state probabilities in large Markov chains and queueing models" IEEE Journal on SAC, Vol. 4 (1986) 920-936.

Dallery, Y., and Y. Frein "A decomposition method for the approximate analysis of closed queueing networks with blocking" Proc. First Int. Workshop on Queueing Networks with Blocking, (H.G. Perros and T. Altiok Eds.) North Holland, 1989.

Dallery, Y., and Y. Frein "On decomposition methods for tandem queueing networks with blocking" Operations Research, Vol. 14 (1993) 386-399.

Frein, Y., and Y. Dallery "Analysis of cyclic queueing networks with finite buffers and blocking before service" Performance Evaluation, Vol. 10 (1989) 197-210.

Gershwin, S. "An efficient decomposition method for the approximate evaluation of tandem queues with finite storage space and blocking" Operations Research, Vol. 35 (1987) 291-305.

Gordon, W.J., and G.F. Newell "Cyclic queueing systems with restricted queues" Oper. Res., Vol. 15 (1967) 286-302.

Hillier, F.S., and R.W. Boling "Finite queues in series with exponential or Erlang service times - a numerical approach" Operations Research, Vol. 15 (1967) 286-303.

Konhein, A.G., and M. Reiser "A queueing model with finite waiting room and blocking" SIAM J. of Computing, Vol. 7 (1978) 210-229.

Kouvatsos, D., and I.U. Awan "Arbitrary closed queueing networks with blocking and multiple job classes" Proc. Third International Workshop on Queueing Networks with Finite Capacity, Bradford, UK, 6-7 July, 1995.

Kouvatsos, D., and S.G. Denazis "Entropy maximized queueing networks with blocking and multiple job classes" Performance Evaluation, Vol. 17 (1993) 189-205.

Kouvatsos, D., and N.P. Xenios "MEM for arbitrary queueing networks with multiple general servers and repetitive-service blocking" Performance Evaluation Vol. 10 (1989) 106-195.

Kritzinger, P., S., van Wyk, and A. Krzesinski "A generalization of Norton's theorem for multiclass queueing networks" Performance Evaluation, Vol. 2 (1982) 98-107.

Jun, K.P., and H.G. Perros "An approximate analysis of open tandem queueing networks with blocking and general service times" Europ. Journal of Operations Research, Vol. 46 (1990) 123-135.

Lee, H.S., and S. M. Pollock "Approximation analysis of open acyclic exponential queueing networks with blocking" Operations Research, Vol. 38 (1990) 1123-1134.

Lee, H.S., A. Bouhchouch, Y. Dallery and Y. Frein "Performance Evaluation of open queueing networks with arbitrary configurations and finite buffers" Proc. Third International Workshop on Queueing Networks with Finite Capacity, Bradford, UK, 6-7 July, 1995.

Liu, X.G., and J.A. Buzacott "A balanced local flow technique for queueing networks with blocking" Proc. First Int. Workshop on Queueing Networks with Blocking (H. Perros and T. Altiok Eds.) North Holland, 1989, 87-104.

Liu, X.G., and J.A. Buzacott "A decomposition related throughput propety of tandem queueing networks with blocking" Queueing Systems, Vol. 13 (1993) 361-383.

Liu, X.G., L. Zwang and J.A. Buzacott "A decomposition method for throughput analysis of cyclic queues with production blocking" in *Queueing Networks with Finite Capacity* (R. O. Onvural and I.F. Akyildiz Eds.) North Holland, 1993, 253-266.

Mishra, S., and S.C. Fang "A maximum entropy optimization approach to tandem queues with generalized blocking" Performance Evaluation, Vol. 30 (1997) 217-241.

Mitra, D., and I. Mitrani " Analysis of a Kanban discipline for cell coordination in production lines I" Management Science, Vol. 36 (1990) 1548-1566.

Mitra, D., and I. Mitrani "Analysis of a Kanban discipline for cell coordination in production lines II: Stochastic demands" Operations Research, Vol. 36 (1992) 807-823.

Onvural, R.O. "Survey of Closed Queueing Networks with Blocking" ACM Computing Surveys, Vol. 22 (1990) 83-121.

Onvural, R.O. Special Issue on Queueing Networks with Finite Capacity, Performance Evaluation, Vol. 17 (1993).

Onvural, R.O., and H.G. Perros "Throughput analysis in cyclic queueing networks with blocking" IEEE Trans. on Software Eng., Vol. 15 (1989) 800-808.

Perros, H.G. "Open queueing networks with blocking" in: *Stochastic Analysis of Computer and Communications Systems* (Takagi Ed.) North Holland, 1989.

Perros, H.G. *Queueing networks with blocking*. Oxford University Press, 1994.

Perros, H.G., and T. Altiok "Approximate analysis of open networks of queues with blocking: tandem configurations" IEEE Trans. on Software Eng., Vol. 12 (1986) 450-461.

Perros, H.G., A. Nilsson and Y.G. Liu "Approximate analysis of product form type queueing networks with blocking and deadlock" Performance Evaluation, Vol. 8 (1988) 19-39.

Perros, H.G., and P.M. Snyder "A computationally efficient approximation algorithm for analyzing queueing networks with blocking" Performance Evaluation, Vol. 9 (1988/89) 217-224.

Shanthikumar, G.J., and D.D. Yao "Monotonicity Properties in Cyclic Queueing Networks with Finite Buffers" in First International Workshop on Queueing Networks with Blocking, (Perros and Altiok Eds), Elsevier Science Publishers, North Holland, 1989.

Suri, R., and G.W. Diehl "A variable buffer size model and its use in analytical closed queueing networks with blocking" Management Sci., Vol.32 (1986) 206-225.

Vantilborgh, H. "Exact aggregation in exponential queueing networks" Journal of the ACM, Vol. 25 (1978) 620-629.

Yao, D.D., and J.A. Buzacott "Modeling a class of state-dependent routing in flexible manufacturing systems" Ann. Oper. Res., Vol. 3 (1985) 153-167.

Yao, D.D., and J.A. Buzacott "Modeling a class of flexible manufacturing systems with reversible routing" Oper. Res., Vol. 35 (1987) 87-93.

PART III

PROPERTIES OF QUEUEING
NETWORKS WITH BLOCKING

7 EQUIVALENCE, INSENSITIVITY AND MONOTONICITY PROPERTIES

Equivalence, insensitivity and monotonicity are important properties of queueing network models that can be used in system performance comparison and evaluation. Insensitivity and equivalence properties provide the basis for comparing the performance of system models with different parameters and with different blocking mechanisms. These results can be applied, for example, to study the impact of the blocking type on system performance, by considering a given set of performance indices and network parameters. In particular, equivalences allow us to define more efficient algorithms to evaluate queueing networks with blocking and to extend the class of models that can be analyzed through analytical methods. Insensitivity properties lead to the identification of the factor that affect system performance and in certain cases it allow a generalization of solution methods. Some important consequence of these properties is that solution methods and algorithms already defined for a certain class of networks could be extended to other classes of networks with different blocking types and/or network parameters. For example, equivalence between networks with and without blocking immediately leads to the extension of efficient computational solution algorithms defined for BCMP networks such as MVA and Convolution algorithm to queueing networks with finite capacity queues. Monotonicity provides insights in the system behavior represented by the queueing network models. It can be applied in parametric analysis to study the impact of various parameters (e.g., system load, buffer dimension) on system performance and possibly to solve optimization problems. It can be used also for bounding analysis. Moreover monotonicity can guide the definition of exact and approximate methods.

In this chapter we consider these properties of queueing networks with finite capacity and blocking. Specifically, in Sections 7.1 through 7.3 we deal with equivalence properties of queueing networks with blocking that arise from the comparison between different models. Equivalence properties are the basis of problem reducibility. Equivalencies include both identity and reducibility relationships and can be defined between networks with and without blocking, between both homogeneous and non-homogeneous networks with different blocking types, and between open and closed networks. It is important to point out that identifying these equivalencies depends on the performance indices involved.

Section 7.1 and 7.2 deal with equivalences in terms of state distribution and average performance indices and in terms of passage time distribution, respectively. Section 7.3 presents some equivalences between the special class of fork and join networks and networks with blocking.

In Section 7.4 we discuss insensitivity properties of queueing networks with blocking, that is how the characteristics of the service requirements affect the network performance.

Finally, monotonicity properties are discussed in Section 7.5 by deriving and studying system performance indices as function of some network parameters.

7.1   EQUIVALENCE PROPERTIES: STATE DISTRIBUTION AND AVERAGE PERFORMANCE INDICES

Equivalences between different network models can be defined by considering various performance indices. Most of the equivalence properties have been defined in terms of identity of the underlying Markov process of the queueing networks, which leads to an identical solution of the stationary state distribution. This often leads also to identity of average performance indices of the queueing networks. Other types of equivalences allow us to define a reducibility relation, i.e. informally to define some relation between the stationary state distribution and between the performance indices. Equivalence properties include relations between open and closed queueing networks with different blocking mechanisms.

Equivalence properties between networks with different blocking types are mostly defined by defining exact transformation functions between state spaces of the associated Markov processes. Such equivalence relations between steady-state queue length distributions allow us to extend solution methods. In particular if two networks with different blocking mechanisms are equivalent, then a product form solution defined for one network can be extended to the other network with the equivalent blocking type. Moreover one can extend or relate insensitivity properties of queueing networks with different blocking mechanisms.

A consequence of the equivalence results is an extension of the class of product form networks with blocking which includes heterogeneous networks, i.e., models where different nodes work under different blocking mechanisms (see definition in Chapter 4). These queueing networks can be used to represent complex systems, for example models of integrated computer-communication systems whose components have different blocking models (e.g., network links, controllers, and computer I/O subsystems can show different types of blocking).

Equivalence relations that can be defined either by identity or by reducibility.

When we define a bijective function between two state spaces of two networks such that the Markov processes are identical, since the meaning of corresponding states may be different, then the performance measures may be not identical. Indeed the network state S definition depends on the blocking type, as discussed in Chapter 4. Moreover the identity of the joint queue length distribution between two networks does not necessarily imply that mean performance indices are identical as well, because their definition depends on the blocking type, as discussed in Chapter 4.

In the following we consider equivalence relationships expressed in terms of state probability π. These equivalences in some cases can be extended to average performance indices such as throughput, utilization, mean queue length and mean response time.

In order to define equivalence properties among different blocking models in terms of steady-state joint queue length distribution, we shall first define two basic relations for two closed or open networks.

First we consider queueing networks with maximum queue length constraints either at node i, B_i, or at node i, chain r, B_{ir}, $1 \le i \le M$, $1 \le r \le R$ (see Section 2.5 for multiclass definition).

Consider two networks with identical parameters (M, C, R, λ, $a_r(\mathbf{m})$, N_r, P, $f_i(n_i)$, μ_{is}, $\forall \mathbf{m}$, $\forall n_i$, $1 \le i \le M$, $1 \le r \le R$, $1 \le s \le C$), with buffer sizes B_i and B_i', or B_{ir} and B_{ir}', $1 \le i \le M$, $1 \le r \le R$, respectively, and different blocking mechanisms X and Y, respectively, where X,Y $\in$ $\mathbb{B}$, where $\mathbb{B}$ is the set of all blocking mechanisms defined in Chapter 2.

Note that, if the two networks are closed then $\lambda = a_r(\mathbf{m}) = 0$, $1 \le r \le R$, $\forall \mathbf{m}$.

Let E and E' be the state spaces of the two networks and let $\pi^X = \{\pi^X(S), S \in E\}$ and $\pi^Y = \{\pi^Y(S'), S' \in E'\}$ denote the corresponding steady-state probability distributions.

We introduce the following relations between the two blocking types X and Y:

- *identity* X=Y: blocking types X and Y are said to be identical if $B_i = B_i'$ or $B_{ir} = B_{ir}'$, $1 \le i \le M$, $1 \le r \le R$, and if E=E', and $\pi^X = \pi^Y$.

- *reducibility* X→Y: blocking type X is said to be reducible to blocking type Y under one of the two following conditions:
  (i) if $B_i = B_i'$ or $B_{ir} = B_{ir}'$, $1 \le i \le M$, $1 \le r \le R$, and if there exists a function f such that $\pi^X = f(\pi^Y)$;
  (ii) if there exist functions g_i or g_{ir} such that $B_i = g_i(B_i')$ or $B_{ir} = g_{ir}(B_{ir}')$, $1 \le i \le M$, $1 \le r \le R$, and if there exists a bijective function f between state spaces E and E' such that steady-state probability is identical for corresponding states, i.e., $\pi^X(S) = \pi^Y(f(S))$, $\forall S \in E$ and $f(S) \in E'$.

7.1.1 Special equivalences

We shall now consider two particular cases of equivalence relationships that can be immediately derived by the definition of the blocking mechanism and some assumptions on the network model.

As discussed in Chapters 2 and 4, according to RS-FD mechanism, when a job finds its destination node full it loops back in the sending node according to the queue discipline. If we assume LIFO discipline, then the job immediately receives a new service without changing its destination. Hence, for exponential queueing networks, because of the memoryless property, this mechanism is equivalent to BBS-SO. In terms of policies behavior, the only difference between the two mechanisms is that the destination of the job is known after (for RS-FD) or before (for BBS-SO) the service. Moreover, when the destination node is full, in a BBS-SO network the service is blocked, while for RS-FD the server continues to serve the job destined to the full node. However, because of the exponential assumption and the independence between routing and service, the Markov chains associated to RS-FD and BBS-SO blocking types are identical, i.e., they have the same state space and state transition rate matrix. In other words,

$$RS\text{-}FD = BBS\text{-}SO \qquad (7.1)$$

for exponential networks with LIFO discipline.

Another equivalence between blocking types can be immediately derived by some assumptions on the network topology. Informally, we state that two blocking types are identical if they describe exactly the same network behavior.
In order to define this case, we first introduce the following definition.

Definition 7.1. Single Destination Property Node. (SDP)
A node i satisfies the Single Destination Property if it has only one destination node j, i.e., if $p_{ij}=1$ and $p_{ik}=0$, $k\neq j$, $1\leq j,k\leq M$.

Figure 7.1 illustrates some basic topologies of open and closed networks. Note that in tandem and in cyclic networks all the nodes satisfy the SDP, and in central server topology networks all the nodes except for node 1 satisfy SDP.

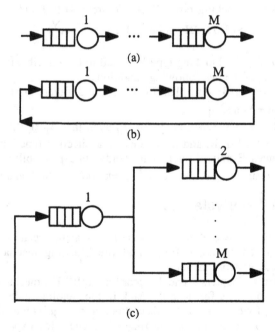

Figure 7.1. (a) Tandem network; (b) Closed cyclic network; (c) Central server network

Consider a node satisfying SDP. Then, it is easy to verify by definition that for such a node the following blocking types are identical:

$$BBS\text{-}SO=BBS\text{-}O \qquad (7.2)$$
$$RS\text{-}RD=RS\text{-}FD \qquad (7.3)$$

These equivalences immediately derive by observing that for the particular node property, the definitions of the two blocking mechanisms become identical.

7.1.2 Equivalences between networks with and without blocking

In this section we consider some equivalences between networks with and without blocking. They allow us to analyse queueing networks with finite capacity by applying standard computational algorithms for queueing networks with infinite capacity, e.g., MVA and Convolution.

By comparing product form solutions of queueing networks with and without blocking one can define a non-blocking network with appropriate parameters such that the stationary state distributions of the two networks are identical. We consider product form networks with blocking introduced in Chapter 5.

Let W denote the network with finite capacity, and W* the network identical to W except for infinite capacity queues and with the following different parameters: load dependent service rate $\mu^*_i f^*_i(n_i)$, and routing matrix P*. Let π^* denote the stationary state distribution of network W*. Single class exponential networks with RS-RD blocking can be proved to be equivalent, in terms of stationary state distribution, to a corresponding network without blocking, as defined in Table 7.1.

Table 7.1. Equivalence between networks with and without blocking

Network with RS-RD blocking	Relationship	Network without blocking parameters
reversible routing solution PF4	$\pi \propto \pi^*$	$\mu_i^* = \mu_i$ $f_i^* (n_i)= f_i (n_i) / b_i (n_i-1)$ $1\leq n_i\leq B_i$ **P* = P**
	$\pi \propto \pi^*$	$\mu_i^* = \mu_i\, h_i$ $f_i^* (n_i)= 1 / b_i (n_i-1)$ $1\leq n_i\leq B_i$ **P* = P**
arbitrary routing solution PF2	$\pi \propto 1 / \pi^*$	$\mu_i^* = \max_j \mu_j$ $f_i^* (n_i)=b_i (n_i-1)$ $1\leq n_i\leq B_i$ $\mathbf{P^*} = \parallel p^*_{ij} \parallel,$ $p^*_{ij}=\mu_j p_{ji}/\mu_i^*\ i\neq j,\ p^*_{ii}=1-\Sigma_{j\neq i}p^*_{ji},\ 1\leq i,j\leq M$

In the first column of Table 7.1 we refer to the product form expressions PF2 and PF4 for networks with RS-RD blocking introduced in Table 5.2 in Chapter 5.

Table 7.1 shows the type of relationship between the two state distributions and the definition of the parameters of the network without blocking. Note that in all the three cases load dependent function $f^*_i(k)$ is defined in the range $1 \leq k \leq B_i$ and it can be any positive arbitrary function for $k > B_i$. In the second case parameter h_i is defined as follows: $h_i = \varepsilon_i \, y_i$ where ε_i is given in PF2 definition in Table 5.2 of Section 5.1 and $y = (y_1, \ldots, y_M)$ is obtained by the solution of $y = y\,A$, where $A = \| a_{ij} \|$, $a_{ij} = p_{ji}$, $j \neq i$, $a_{ii} = 1 - \Sigma_{j \neq i}\, a_{ij}$, $1 \leq i,j \leq M$.

These equivalences are obtained by comparing the product form solution of the two networks that yield the same results by an appropriate choice of the network parameters. In particular note that when the blocking function b_i are 0-1 function (i.e., $b_i(n_i) = 1$ for $0 \leq n_i < B_i$, $b_i(B_i) = 0$, $1 \leq i \leq M$), then the first two equivalence in Table 7.1 state that the network with blocking has the same solution as the network without blocking, up to a normalizing constant on the reduced state space. The first case of reversible routing networks is also related to the property of the underlying Markov processes of the two networks: the process of the network with blocking is the truncated Markov process of the network without blocking. Since the process of the network without blocking has product form solution, then also the truncated process has the same solution normalized in the reduced state space. The second and third equivalences in Table 7.1 for arbitrary routing network are less intuitive and they derive from the definition of product form solution for the two networks and by using some algebra. The third equivalence defines an inverse proportional relation between the state probability of the two networks. This is obtained by defining from the network with blocking a dual network without blocking with routing matrix P^*. A detailed proof of these equivalences can be found in the references given in Section 7.6.

Moreover we observe that since the product form expression for queueing networks with finite capacity given by formula (5.1) has a similar structure as the product form expression of networks with infinite capacity queues, these equivalence properties could be extended to other cases of product form networks with blocking, including multiclass networks with different types of nodes. In other words, by comparing the product form solution of networks with blocking, multiple classes of customers and various network topologies, as described in Chapter 5, with the corresponding product form solution of networks without blocking, one could extend these equivalences to other cases.

7.1.3 Equivalences between different blocking types

Equivalences between networks with different blocking types have been identified both for open and closed networks. They include both identity relationship and reducibility. As introduced above, identity states that the state definition and the state distributions of the two networks are identical, while reducibility defines a correspondence between the two distributions. Most of the reducibility between networks can be defined by modifying the buffer capacities, i.e., they are defined according to condition (ii).

Let B_i^X denote the buffer capacity when node i works under blocking type X and π^X the stationary state distribution of the homogeneous network with blocking type X.

Tables 7.2, 7.3 and 7.4 show equivalences of state distribution and, in some cases, of average performance indices between some blocking types for certain networks with various topology.

Note that for the sake of simplicity in these tables U_i and X_i respectively denote the *effective* utilization and the *effective* throughput of node i for each blocking type, as defined in chapter 1.

Tables 7.2 and 7.3 show equivalences for closed and open networks, respectively and special topology. Table 7.4 shows equivalences for both closed and open networks with more general topology.

Closed networks with special topology are two node networks, cyclic and central server topologies. Node 1, as illustrated in figure 7.1, denotes the central node in central server networks. The tables show the condition under which the equivalence properties between different blocking type hold. They include identity and reducibility relations for each special topology networks and for the set of performance indices that satisfy the equivalence.

Two node networks

First consider a two-node network. In order to avoid deadlock we assume that $N<B_1+B_2$. Moreover BBS-SO is identical to BBS-SNO when $N<B_1+B_2-1$, because the blocked node has always room to keep the server free.

Equivalences (7.2) and (7.3) hold for two-node and cyclic networks without additional assumptions, as illustrated in Table 7.2. They are expressed in terms of state probability π and average performance indices. Hereafter we use BBS and RS to denote the set of blocking mechanisms when equivalences hold for all the subtypes (SO, SNO, O and RD, FD, respectively).

BBS and RS mechanism are equivalent for multiclass two-node networks with BCMP type nodes and class independent capacities, i.e. assumption (I). The proof of this equivalence is similar to the proof for single-class exponential networks, but with a cumbersome notation. This property basically derives from the mechanism definition and the special network topology. Moreover under identical assumptions BAS is reducible to BBS mechanism provided that node capacities are augmented by 1. It is easy to be convinced that the behavior of the two networks is identical, as illustrated by the following example.

Example 7.1. For the sake of simplicity let us consider an exponential closed network with two single server nodes and a single class of users. The nodes work under BAS mechanism and have queue capacity $B_1=B_2=3$. Let N=5 be the network population. According to the state definition $S_i=(n_i,s_i,m_i)$ given in Chapter 4 the state space is defined as follows:

$$E_{BAS} = \{((3, 0,[2]), (2,1,\varnothing)), ((3, 0,\varnothing), (2,0,\varnothing)), ((2, 0,\varnothing), (3,0,\varnothing)), ((2,1,\varnothing), (3, 0,[1]))\}$$

The underlying Markov process is a birth-death process with rates μ_1 and μ_2 and the state transition diagram is shown in figure 7.2 (a).

Table 7.2. Equivalence between closed networks with different blocking types

Network topology	Performance indices	Blocking types	Assumptions
two-node	π U_i,X_i,L_i,T_i	BBS-SO=BBS-O RS-RD=RS-FD	
		BBS-SO=BBS-SNO	$N<B_1+B_2-1$
	π	BBS=RS	(I) : multiclass networks BCMP type nodes class independent capacities
		BBS→BAS	assumption (I) and with $B_i^{BBS}=B_i^{BAS}+1$, $1\leq i\leq M$
	π U_i,X_i,L_i,T_i	BBS=RS	(II) : single class networks exponential nodes load independent service rates
	π U_i,X_i	BBS→BAS	assumption (II) and with $B_i^{BBS}=B_i^{BAS}+1$, $\forall i$
cyclic	π U_i,X_i,L_i,T_i	BBS-SO=BBS-O RS-RD=RS-FD	
		BBS-SO=BBS-SNO	$N<\min\{B_i+B_j : p_{ij}>0\}$
		BBS-SO=BBS-O=RS	assumption (II)
	π U_i, X_i	BBS-SO→BAS	assumption (II) and with $B_i^{BBS\text{-}SO}=B_i^{BAS}+1$, $\forall i$
central server	π U_i,X_i,L_i,T_i	BBS-SO=BBS-SNO =BBS-O RS-RD=RS-FD	if only $B_1<\infty$ and $B_i=\infty$, $2\leq i\leq M$
		BBS=RS	if only $B_1<\infty$ and $B_i=\infty$, $2\leq i\leq M$, and assumption (II)
		BBS-SO=BBS-SNO	if $B_1=\infty$, $B_i<\infty$, $2\leq i\leq M$
		BBS-SO=BBS-SNO =RS-FD	if $B_1=\infty$, $B_i<\infty$, $2\leq i\leq M$ assumption (II) and LIFO
	π U_i, X_i	BBS-O→BAS	assumption (II) and if $B_1=\infty$, $B_i<\infty$, and $B_i^{BBS\text{-}O}=B_i^{BAS}+1$, $2\leq i\leq M$

Consider state $S=((3, 0,\varnothing), (2,0,\varnothing))$ where 3 customers are in node 1 which is full, 2 customers are in node 2 and both nodes are active. According to the blocking mechanism when node 2 ends the service it becomes blocked and the state is: $S'=((3, 0,[2]), (2,1,\varnothing))$. Then in state S' only node 1 is active and when it ends the service, according to the simultaneous transition assumption (see BAS definition in Section 2.2) the state becomes again $S=((3, 0,\varnothing), (2,0,\varnothing))$.

Let us now consider the same network but with BBS mechanism and augmented queue capacities $B_1=B_2=4$. Since the blocking sub-types of BBS are equivalent, we can use the simpler node state notation $S_i=(n_i)$ as for the BBS-O mechanism. The network state space is given by:

$$E_{BBS} = \{(4, 1), (3, 2), (2, 3), (1, 4)\}$$

The associated Markov process is a birth-death process with rates μ_1 and μ_2 and the state transition diagram is shown in figure 7.2 (b). In states (3,2) and (2,3) both nodes are active, while in states (4,1) and (1,4) nodes 2 and 1 are blocked, respectively.

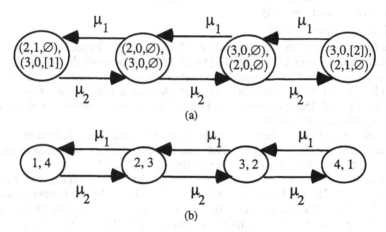

(a)

(b)

Figure 7.2. Markov state transition diagram: (a) network with BAS blocking; (b) network with BBS blocking and augmented queue capacities

Let us define the following bijective function f between networks' state spaces E_{BBS} and E_{BAS}:

$$f((4, 1)) = ((3, 0,[2]), (2,1,\varnothing))$$
$$f((3, 2)) = ((3, 0,\varnothing), (2,0,\varnothing))$$
$$f((2, 3)) = ((2, 0,\varnothing), (3,0,\varnothing))$$
$$f((1, 4)) = ((2,1,\varnothing), (3, 0,[1]))$$

It is easy to be show that the steady-state probability of the two models are identical for corresponding states, i.e., $\pi^{BBS}(S)=\pi^{BAS}(f(S))$, $\forall S \in E_{BBS}$ and $f(S) \in E_{BAS}$.

Cyclic networks

Equivalences (7.2) and (7.3) hold for cyclic networks without additional assumptions. The equivalences for two-node networks under assumption (I) are expressed only in terms of the state probability π. If we further restrict to single class exponential networks with load independent service rates (assumption (II)), these equivalences extend to cyclic networks. They are expressed not only in terms of the state probability π, but also for all the average performance indices for the identity BBS-SO=RS and for utilization and throughput for reducibility between BBS and BAS.

Similarly to the case of two node networks, we can prove for cyclic network equivalences between BBS-SO and RS mechanisms and between BBS-SO and BBS-SNO, with the further assumption that guarantees that a blocked node has always room to keep the server free.

Central server networks

Let us consider a central server network for two particular cases:
 1. $B_1 < \infty$ and $B_i = \infty$, $2 \leq i \leq M$
 2. $B_1 = \infty$, $B_i < \infty$, $2 \leq i \leq M$.
In the first case, only the central node has finite capacity, hence due to the special topology equivalences (7.2) and (7.3) hold. We also observe that BBS-SO=BBS-SNO because all the node except the central one has infinite capacity and a blocked node i has always room to keep its server free. Moreover under assumption (II) BBS is identical to RS. These equivalences hold for the state probability π and for all the average performance indices.

In the second case, when each node except the central one has finite capacity, only the central node may be blocked. Since node 1 has always room to keep its server free then we have the equivalence BBS-SO=BBS-SNO.

Note that the central node does not satisfy the SDP property, consequently equivalences (7.2) and (7.3) do not hold and BBS-SO and RS-RD behave differently from BBS-O and RS-FD, respectively. However, for the special case of assumption (II) and node 1 with LIFO discipline we obtain BBS-SO=RS-FD, as explained in Section 7.1.1.

Finally, BAS mechanism is reducible to BBS-O mechanism with node capacities B_i, $2 \leq i \leq M$, augmented by 1. Analogously to two node networks we observe that network behavior for the two mechanisms are identical. Indeed, according to BAS mechanism, when a node i completes a service for a full node j, node i is blocked until node j ends a service. At completion service time in node j we observe a simultaneous transition and the job waiting in i is immediately transferred into destination node j. This is equivalent to have an extra-position in the buffer of destination node j and blocking node i after the end of the service of the job destined to j. This is what happens according to BBS-O mechanism with augmented buffer capacities.

Note that in this case BBS-SO mechanism is not equivalent to the other blocking type. Indeed, BBS-SO has a greater state space than the other ones, since the BBS-SO process state includes component NS_1 for each state with $n_1 \neq 0$ to denote the destination of the job currently in service (see chapter 4, Section 4.1.4).

Table 7.3. Equivalence between open networks with different blocking types

Network topology	Performance indices	Blocking types	Assumptions
tandem	π U_i, X_i, L_i, T_i	BBS-SO=BBS-O RS-RD=RS-FD	
		BBS-SO=BBS-SNO	$N < \min_{1 \le i < M} \{B_i + B_{i+1}\}$
		BBS-SO=BBS-O=RS	assumption (II)
	π U_i, X_i	BBS-SO$\rightarrow$BAS	assumption (II) and with $B_i BBS\text{-}SO = B_i BAS + 1$, $2 \le i \le M$
split	π U_i, X_i, L_i, T_i	BBS-SO=BBS-SNO	$N < \min_{2 \le i \le M} \{B_1 + B_i\}$
		BBS-SO=RS-FD	assumption (II) and LIFO
	π U_i, X_i	BBS-O$\rightarrow$BAS	assumption (II) and with $B_i BBS\text{-}O = B_i BAS + 1$, $2 \le i \le M$
merge	π U_i, X_i, L_i, T_i	BBS-SO=BBS-O RS-RD=RS-FD	
		BBS-SO=BBS-SNO	$N < \min_{2 \le i \le M} \{B_1 + B_i\}$
		BBS-SO=BBS-O=RS	assumption (II)

Tandem, split and merge topology open networks

Table 7.3 presents equivalence properties for some open networks, that are tandem and two special cases of open networks denoted as split and merge topology networks, illustrated in figure 7.3.

Split topology, shown in figure 7.3 (b), can be defined as follows: $p_{01}=1$, $p_{0i}=0$, $2 \le i \le M$, $p_{ij}>0$ for $i=1$ and $2 \le j \le M$, $p_{ij}=0$ otherwise, $p_{10}=0$, $p_{i0}=1$ for $2 \le i \le M$, i.e., an external arrival enters the network only at node 1, from which it can go to nodes 2,..., M and from which it eventually exits the network.

Merge topology, shown in figure 7.3 (c), is defined as follows: $p_{01}=0$, $p_{0i}>0$, $2 \le i \le M$, $p_{ij}>0$ for $2 \le i \le M$ and $j=1$, $p_{ij}=0$ otherwise, $p_{10}=1$, $p_{i0}=0$ for $2 \le i \le M$, i.e., an external arrival enters in any of nodes 2,..., M and then it goes to node 1 from which it leaves the network.

Both tandem and merge topologies satisfy for each node the SDP property defined in Section 7.1.1. As a consequence, the topological equivalences (7.2) and (7.3) hold. These equivalence relationships are expressed in terms of both state probability π and average performance indices.

All the equivalences shown in Table 7.3 have a similar counterpart in the Table 7.2 for the corresponding closed topologies and we do not explain them further.

Reversible routing and arbitrary topology networks

Table 7.4 refers to both open and closed networks for the two more general cases of reversible routing and arbitrary topology networks. Various particular equivalences have been derived under special assumptions.

In the Table we use the A-type node definition given in Chapter 5. A node is said to be A-type in a single class network if it has an arbitrary service time

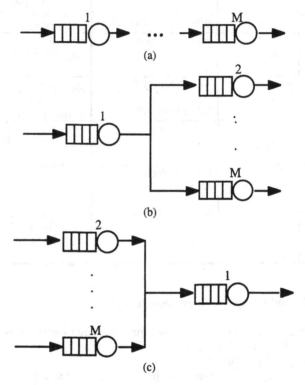

Figure 7.3. (a) Tandem network; (b) Split topology network; (c) Merge topology network

distribution and a symmetric scheduling discipline or exponential service time distribution and arbitrary scheduling.

For reversible routing network the only equivalence relates RS-RD and Stop blocking for a single class closed or open network with A-type nodes and load independent service. Note that this is a case of network with product form solution.

For arbitrary topology networks Table 7.4 shows several special cases.

The first case derives from the special equivalence relationships introduced in Section 7.1.1 based on topological properties (Single Destination Property) or on exponential assumption that may hold also for arbitrary topology. In particular, if each node with finite capacity satisfies the SDP then the identity relationships (7.2) and (7.3) hold.

Table 7.4.
Equivalence between networks with more general topology and different blocking types

Network topology	Performance indices	Blocking types	Assumptions
reversible routing	π	RS-RD=Stop	single class closed/open networks A-type nodes load independent service
arbitrary routing	π U_i, X_i, L_i, T_i	BBS-SO=BBS-O RS-RD=RS-FD	SDP for each node with finite capacity
		BBS-SO=RS-FD	assumption (II) and LIFO
		BBS-SO= =BBS-O= =RS-RD=RS-FD	assumption (II) and SDP for each node with finite capacity
		BBS-SO= =BBS-SNO	$N < \min \{B_i+B_j$ for i,j: $(p_{ij}>0$ and $p_{ji}=0)$ or $B_i+B_j-1)$ for i,j: $(p_{ij}>0$ and $p_{ji}>0)\}$
	π	Stop=Recirculate	multiclass open Jackson networks with class type fixed
		Stop→BBS-O (open) (closed)	single class Jackson networks

The second equivalence holds regardless the network topology between RS-FD and BBS-SO for exponential single class queueing networks, with LIFO discipline and routing independent of service time. This is the third row in Table 7.2

Moreover BBS (subtypes SO and O) is identical to RS mechanism for single class exponential networks with load independent service rates and each node with finite capacity satisfying the SDP property.

In the particular case of network population

$$N < \min \{B_i+B_j \text{ for i,j} : (p_{ij}>0 \text{ and } p_{ji}=0) \text{ or } B_i+B_j-1 \text{ for i,j} : (p_{ij}>0 \text{ and } p_{ji}>0)\}$$

BBS-SO is identical to BBS-SNO since the blocked node has always room to keep free the server. This constraint on the maximum population does not allow two nodes to be full at the same time. Note that for BBS-SNO the second condition in the expression above is necessary to guarantee deadlock freedom.

For all these first four cases of arbitrary networks, the equivalence relationships are expressed in terms of both state probability π and average performance indices.

Another identity holds for arbitrary topology networks in terms of state probability π. Stop and Recirculate blocking mechanisms are identical for open networks with FCFS exponential nodes, Poisson arrivals and multiple job classes but with the job class-type fixed throughout its residence in the system. Note that this is a case of product form network.

The last equivalence reported in Table 7.4 is a special case which relates an open network with M nodes and Stop blocking, with a closed network, with an additional node with appropriate parameters and BBS-O blocking. In particular consider an open Jackson network whose population n satisfies constraints $L \leq n \leq U$, with L, $U \geq 0$, and with node finite capacities B_i, $1 \leq i \leq M$. The arrival rate function a(n) and the departure blocking function d(n), when there are n jobs are in the network, satisfy the following constraints:

$$a(U)=0 \text{ if } U<\infty \text{ and } a(n)>0, \text{ for } L \leq n<U,$$
$$d(L)=0 \text{ if } L>0 \text{ and } d(n)=1, \text{ for } L<n \leq U.$$

It is known that any open queueing network with finite queues and Poisson arrivals can be exactly analyzed as a closed queueing network. We call this closed queueing network the "analogue" closed network of the open network. Specifically, the closed network is defined by adding a node, denoted by 0, with FCFS discipline and the following characteristics:

load dependent service rate: $\mu_0(n_0)=a(n)$
finite capacity: $B_0=U-L$
blocking function: $b_0(n_0)=d(n)$

where $n_0=U-n$, $0 \leq n_0 \leq B_0$.

Node 0 represents the external population of the open network and n is the total population of subnetwork $\{1,2,...,M\}$: $n=\Sigma_{1 \leq i \leq M} n_i$, $L \leq n \leq U$. The total population in analogue closed network the is N=U.

If node 0 works under BBS-O blocking, by blocking type definition, one can immediately derive that the closed network behaves like the open Jackson network under STOP mechanism. Hence it is the analogue closed network of the open one. The open and the corresponding closed networks are shown in figure 7.4.

Hence the following correspondence between state distributions holds:

$$\pi^{Stop}(S)=\pi^{BBS-O}(n_0,S)$$

for each state S of the open network.

Heterogeneous networks

Finally, note that Tables 7.2, 7.3 and 7.4 show equivalence properties between homogeneous networks with different blocking types and the corresponding conditions and assumptions, including network characteristics and special conditions on system parameters.

We point out that all these equivalence properties may be used to define heterogeneous networks, where different nodes can work under different blocking types. As a consequence, we can extend the presented equivalences to relate the performance indices of two heterogeneous networks, by combining the various equivalences illustrated in Tables 7.1 through 7.4. In addition a heterogeneous network can be also related to a homogeneous network with the equivalent blocking type.

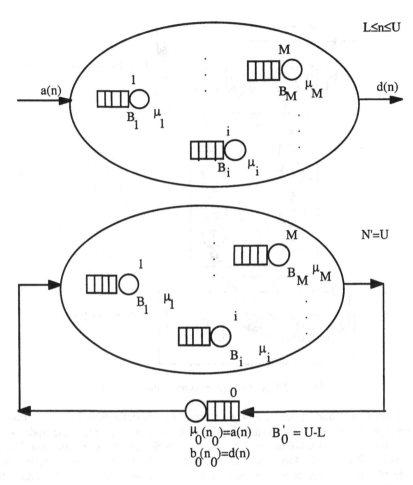

Figure 7.4. An open Jackson network and the analogue closed network

Remark. Note that heterogeneous networks where service centers work under different and equivalent blocking mechanisms are also equivalent to homogeneous networks with one of the considered blocking types.

Example 7.2. Consider the heterogeneous illustrated in figure 7.5. It is a simple model (possibly an aggregated one) of distributed systems where two computer systems (nodes 1, 2, 5 and 6) are interconnected by a communication network (nodes 3, 4, 7 and 8). The network is analyzed with the following parameters: M=8, N=4, $\mu_i=1$, $B_i=2$ for i=1, 2, 5, 6 and $\mu_i=10$, $B_i=1$ for i=3, 4, 7, 8 and routing matrix P defined as follows: $p_{21}=p_{34}=p_{46}=p_{65}=p_{78}=p_{82}=1$, $p_{13}=p_{57}=0.3$, $p_{12}=p_{56}=0.7$ and $p_{ij}=0$ otherwise. Note that nodes 1, 4, 5 and 8 satisfy condition (B) defined in chapter

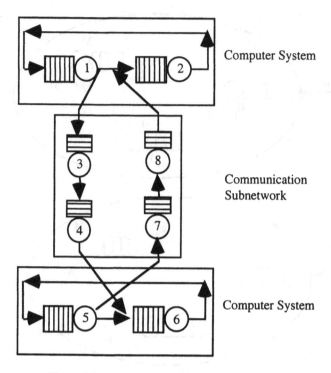

Figure 7.5. A computer communication network model

5, definition 5.4, and nodes 2, 3, 4, 6, 7 and 8 satisfy definition 7.1. Since nodes 2 and 6 represent disk I/O subsystem, BAS blocking is considered, while nodes 4 and 8, that represent communication links, have BBS-SO blocking mechanism. Finally we assume RS-RD blocking for nodes 1, 3, 5 and 7. By using equivalence results illustrated above this heterogeneous network is reducible to an homogeneous network with RS-RD blocking type with the same parameters except for node 1 and 5 new capacities $B_1^{RS-RD} = B_5^{RS-RD} = 3$. We can define the bijective function f between state spaces of the heterogeneous and the homogeneous network E_H and E_{RS-RD} such that steady-state probability is identical for corresponding states, i.e., $\pi^H(S) = \pi^{RS-RD}(f(S))$, $\forall S \in E_H$ and $f(S) \in E_{RS-RD}$. For example some correspondent states of the two networks are the following:

$$S=(3,1,0,0,0,0,0,0), f(S)=(2,(2,1),0,0,0,0,0,0)$$
$$S=(3,0,1,0,0,0,0,0), f(S)=(2,(1,1),1,0,0,0,0,0)$$
$$S=(1,1,0,0,1,1,0,0), f(S)=(1,(1,1),0,0,1,(1,1),0,0)$$
$$S=(0,2,0,0,2,0,0,0), f(S)=(0,(2,1),0,0,2,0,0,0)$$
$$S=(0,0,0,0,3,1,0,0), f(S)=(0,0,0,0,2,(2,1),0,0)$$

We obtain the same node utilization and throughput but different mean response time in the two networks. For example for node 1 we obtain the same utilization $U_1=0.546$ but different mean response time $T^H_1=1.8554$ and $T^{RS-RD}_1=1.9678$.

Note that if we consider RS-RD blocking at node 2 and 6 and if the network has exponential service time and constant service rates, then the model in figure 7.5 gives an example of non-homogeneous product form network with solution PF2, as defined in chapter 5, Table 5.1.

7.2   EQUIVALENCE PROPERTIES: PASSAGE TIME DISTRIBUTION

In this section we present some equivalence results in term of passage time distribution.

Equivalences between networks defined in terms of joint queue length distribution and average performance indices presented in the previous section do not necessarily lead to equivalence in terms of passage time distributions.

The evaluation of the passage time, cycle time and network response time distribution in general is a difficult problem.

Some equivalence results have been obtained in terms of cycle time distributions for cyclic networks with BAS and BBS-SO blocking types.

The extension of such results to other networks with different parameters including blocking type, service distribution and routing topology is an open issue.

Consider a single class two-node cyclic exponential network with N customers, finite capacities B_1 and B_2 and either BBS-SO or BAS blocking. Let $f_{N,B_1,B_2}(t)$ denote the density function of the cycle time. The following equivalence property can be proved:

Theorem 7.1

Consider two single class cyclic networks with two exponential nodes, N customers, service rates μ_i, i=1,2, BBS-SO or BAS blocking and finite capacities B_i and B_i', respectively, for i=1,2. If

$$B_1+B_2 = B_1'+B_2'$$

then the two networks are equivalent in terms of cycle time distribution, i.e.:

$$f_{N,B_1,B_2}(t) = f_{N,B'_1,B'_2}(t).$$

In other words this equivalence states that the distribution of the cycle time does not depend on the single buffer size of each node, but on the total buffer capacity of the network. The extension of these equivalencies to queueing networks with a more general topology and different blocking types is an open issue.

Remark. Note that since the two networks have the same number of customers but different capacities they are also equivalent in terms of throughput, but they are not equivalent in terms of joint queue length distribution and other average performance indices (mean response time, utilization and mean queue length).

Example 7.3. Consider a two node cyclic network with N=10 customers, service rates $\mu_1=1$, $\mu_2=2$, BBS-SO blocking and finite capacities B_1 and B_2. By theorem 7.1 we observe the same cycle time distribution for each pair of finite capacities B_1 and B_2 provided that the total capacity B_1+B_2 is constant, that is $\forall(B_1, B_2): B_1+B_2 =B=1,...,10$. For example the two networks with capacities $B_1=2$, $B_2 =7$ and $B'_1=5$, $B'_2 =4$ are equivalent in terms of passage time distribution. Note that the two networks have the same throughput, but they are not equivalent in terms of queue length distribution, node utilization and mean response time.

7.3 EQUIVALENCE PROPERTIES: FORK/JOIN NETWORKS

Queueing networks with fork and join operations and finite capacity queues are used as performance evaluation model of a class of discrete event systems such as parallel processing computer systems and manufacturing systems. For example in parallel processing computer systems forking occurs when a customer or job is split into a set of sub-customers or tasks that can run in parallel on different processing units and joining occurs when a set of separate jobs or tasks synchronize their executions. In manufacturing systems forking is the process of splitting a single product, possibly an assembled one, into a set of parts and joining is the assembly process of a set of parts to form a single product.

  Different queueing network models with fork and join can be defined. Moreover the definition of the blocking mechanisms for such models is not immediate.

  We consider a special model called Fork/Join Queueing Networks with Blocking for which several properties have been proved, including equivalences, monotonicity, reversibility, symmetry and concavity of some performance indices. In a fork node a customer is split into a set of customers that on departing from the server go to different queues. In a join node the server requires a set of customers from different queues before it can start the service and this represents synchronization among the arriving customers. At the completion of the join operation the customer exit the server. Let us define the class of Fork/Join Queueing Networks with Blocking. In this Section we present some equivalence properties that hold for such networks.

  A Fork/Join Queueing Networks with Blocking (FJQNB) is a queueing network consisting of a set of servers and a set of buffers such that each buffer has exactly one upstream server and one downstream server. On the other hand, each server may have several input and/or output buffers. Each buffer has finite capacity. Each server is allowed to start service when there is at least one customer in each of its input buffers and space for at least one customer in each of its output buffers. The model is open and there are sources and sinks that are special servers without upstream and downstream buffer, respectively. We assume general service times, in particular distribution that they are from jointly stationary, ergodic and reversible sequences of integrable random variables (see the references reported in Section 7.6 for further details). We assume the BBS-SO blocking, i.e. the server is blocked if any of its downstream buffers is full.

In FJQNB's one can define two types of servers depending on whether the server has space to accomodate all the jobs in service, that is a job for each of its upstream buffers. If the server has space to accomodate jobs one can define various blocking mechanims as well as loading and unloading policies. When such a server performs an operation it consumes one job for each upstream buffer and places one job in each of its downstream buffers. This server cannot begin the service activity unless all the jobs, one from each of it upstream buffers, have been loaded onto the server. The server cannot load a new job before all the jobs produced during the previous service activity have been unloaded. For example different loading and unloading policies can be defined by considering independent or simultaneous loading/unloading of the jobs.

Note that this model is different from the class of queueing networks with blocking that we have introduced and studied in this book. The servers and the buffers or queues are considered separated and they do not form the service centers of the classical queueing networks. The topology of such networks is specified by listing the upstream and downstream servers that are connected to each buffer. Note that in FJQNB we do not define a routing probability matrix. Then the structure of the FJQNB can be represented by a FJQNB graph whose nodes are the servers and the buffers. Simple examples of an open FJQNB is illustrated in figure 7.6, where the servers are circles and the buffers are rectangles.

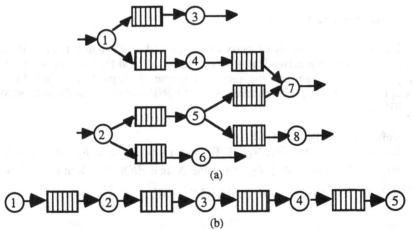

Figure 7.6. Examples of Fork/Join queueing network with Blocking: (a) an open FJQNB, (b) an open tandem queueing network as a FJQNB

One can also prove that any FJQNB can be transformed into an equivalent so-called canonical FJQNB where there is at most one buffer between any pair of servers. In this case the FJQNB can be represented by a directed graph whose nodes are the servers and whose edges are the buffers.

A sufficient condition for the existence of the stationary solution of the queue probability distribution in open FJQNB is that they are acyclic, not taking into account the direction of flow in each buffer (see the references given in Section 7.6 for further details). Figure 7.6 (a) shows an example of such an open FJQNB.

Queueing networks models with tandem or cyclic topology are special cases of FJQNB. Figure 7.6 (b) shows a tandem queueing network seen as a particular FJQNB, where server 1 is the external arrival source and server 5 is a sink. Several equivalence results and properties have been first proved for tandem and/or cyclic topology queueing networks and can be generalized to FJQNB.

Let V_s and V_b denote respectively the set of n_s servers and n_b buffers of the FJQNB. Let E be the set of directed edges that denote the flow of customers from servers to buffers and viceversa, where

$$E \subset V_s \times V_b + V_b \times V_s$$
$$\forall k \in V_b : |\{(i,k) \in E : i \in V_s\}| = 1 \text{ and } |\{(k,i) \in E : i \in V_s\}| = 1$$

that is each buffer has one incoming and outgoing edge. Let B_i denote the queue capacity of buffer i. Let $\mathbf{B} = (B_1, ..., B_{n_b})$.

Then a FJQNB is defined as $W = (V_s, V_b, E, \mathbf{B})$.

The behaviour of the FJQNB in general depends on an initial state, also called initial marking. Let $M = (M_1, ..., M_{n_b})$, $0 \leq M_i \leq B_i$, denote the initial marking of the FJQNB, where M_i denote the number of customers in buffer i at initial time t=0, $\forall i$. Note that $B_i - M_i$ is the number of holes in buffer i at time t=0. Similarly, let $M(t) = (M_1(t), ..., M_{n_b}(t))$ denote the marking at time t.

Reversibility and duality

Throughput reversibility property can be proved for tandem networks. It states that an open queueing network with tandem topology and the same network where the queues are reversed have the same throughput. Such property is based on the concept of duality and can be generalized for FJQNB. Let us define the dual network of a FJQNB.

Definition 7.2. Dual FJQNB.

Given a FJQNB $W = (V_s, V_b, E, \mathbf{B})$ with initial marking $\mathbf{M}$, let Δ denote an arbitrary set of buffers, $\Delta \subset V_b$. Then the Δ-dual FJQNB is defined as follows: $W^D = (V_s, V_b, E^D, \mathbf{B})$ with initial marking $M^D = (M_1^D, ..., M_{n_b}^D)$, where

$$E^D = E - \{(i,k) \in E : k \in \Delta\} + \{(k,i) : (i,k) \in E : k \in \Delta\}$$
$$- \{(k,j) \in E : k \in \Delta\} + \{(j,k) : (k,j) \in E : k \in \Delta\}$$
$$M_i^D = B_i - M_i \text{ for } i \in \Delta, \; M_i^D = M_i \text{ otherwise}$$

Note that if W is without cycles then any Δ-dual network of W is also without cycles. If $\Delta = V_b$ we have the *full dual* network of W. This full dual network is obtained by reversing the customer flow through all the buffers and switching all the initial markings of the original network with the holes.

One can define a bijective function between the customers in the buffers within Δ of the original network W and the holes within the same buffers in the Δ-dual network W^D and viceversa. Using duality one can prove the following theorem. Let $X(W)$ denote the throughput of network W.

Theorem 7.2

If the networks W and W^D have the same service times and initial timing condition, then:

(i)   for all $t \geq 0$, $M_i^D(t) = B_i - M_i(t)$ for $i \in \Delta$, $M_i^D(t) = M_i(t)$ otherwise.

(ii)   Whenever the throughput exists for network W, then the throughput exists for network W^D and is identical to that of W, i.e.:

$$X(W) = X(W^D).$$

(iii)   Whenever the stationary distribution of the system marking exists for network W, the stationary distribution of the system marking exists for network W^D and is related with that of W as in (i).

Another equivalence property can be proved between the class of Fork/Join Queueing Networks with Blocking and a special case of Petri Net called Strongly Connected Markov Graphs.

Other equivalences in terms of throughputs can be derived between FJQNB's with different initial timing conditions and marking. The interested reader may refer to the references in Section 7.6.

We shall now define the reversed network of a FJQNB. Consider a canonical FJQNB where there is at most one buffer between any pair of servers, which is represented by a directed graph whose nodes are the servers and whose edges are the buffers. Then we define $W = (V_S, E, \mathbf{B})$.

Definition 7.3. Reverse of FJQNB.

Given a canonical FJQNB $W = (V_S, E, \mathbf{B})$ with initial marking $\mathbf{M}$, the reverse of W is defined as follows: $W^R = (V_S, E^R, \mathbf{B}^R)$ with initial marking $\mathbf{M}^R$, where

$$E^R = \{(i, j) : (j, i) \in E\}$$
$$B_{i,j}^R = B_{j,i}, \quad M_{i,j}^R = M_{j,i} \text{ for } (j, i) \in E$$

The following results can be proved.

Theorem 7.3

A FJQNB W is deadlock-free if and only if its reverse W^R is deadlock-free.

Theorem 7.4

Given an arbitrary deadlock-free FJQNB W and its reverse W^R with the same joint distribution of the sequence of service times, if the service times form jointly reversible, stationary and ergodic sequence of integrable random variables, then the two networks have the same throughput, i.e.

$$X(W) = X(W^R).$$

Note that in particular theorem 7.4 is satisfied when the service times are mutually independent and are independent and identically distributed (i.i.d.) at each server.

Symmetry

By combining reversibility and duality property one can prove that the throughput of a FJQNB with a given initial marking is identical to that of the same network with symmetrical marking, as stated by the following result.

Definition 7.4. Symmetrical of FJQNB.

Given a canonical FJQNB $W=(V_S, E, B)$ with initial marking M, the symmetrical FJQNB of W is defined as follows: $W^S=(V_S, E, B)$ with initial marking $M^S=(B-M)$.

Theorem 7.5

Given an arbitrary deadlock-free FJQNB W, if the service times form jointly reversible, stationary and ergodic sequence of integrable random variables, then the throughput of W with initial marking M is the same as with initial marking $(B-M)$, i.e.

$$X(W) = X(W^S).$$

This result for the special case of closed cyclic networks provides the following result of practical interest. Given a network W with population N, let $X_W(N)$ denote the network throughput.

Corollary 7.1

Given a *closed cyclic* FJQNB, if the service times form jointly reversible, stationary and ergodic sequence of integrable random variables, and the sequences of service times at different servers are mutually independent, then the throughput of the network with total buffer capacity B and N customers is the same as with total buffer capacity B and B-N customers, i.e.

$$X_W(N) = X_W(B-N), \quad 1 \leq N \leq B-1.$$

A similar property holds for a class of closed FJQNB called series-parallel, where servers and buffers have a particular connection, as stated by the following theorem. A closed series-parallel FJQNB is called *uniform* if the sum of the buffer capacities in all circuits is identical, and the total number of customers in all circuits is also identical.

Theorem 7.6

Given an arbitrary *uniform closed series-parallel* FJQNB W, if the service times form jointly reversible, stationary and ergodic sequence of integrable random variables and the sequences of service times at different servers are mutually independent, then the throughput of W of the network with total buffer capacity B and N customers is the same as with total buffer capacity B and B-N customers, i.e.

$$X_W (N) = X_W (B-N), \ 1 \le N \le B-1).$$

Concavity

Finally, one can prove the throughput of a FJQNB is a concave function with respect to both the buffer capacities and the initial marking, as stated by the following result. The service times are assumed to belong to the class of PERT distributions which include the Erlang distribution (see the references for further details).

Definition 7.5. PERT type distribution.

A stochastic PERT graph is a weighted directed acyclic graph where the weights are random variables associated with the nodes. The weight of the critical path of the stochastic PERT graph is the maximum of the weights of all the paths in the graph.

Random variable X has a PERT type distribution if X can be expressed as the weigth of a critical path of a stochastic PERT graph where the weights of the nodes are mutually independent exponential random variables.

Theorem 7.7

Given a canonical FJQNB W, if the service times form mutually independent sequences of i.i.d. random variables having PERT distribution, then the throughput of W is a concave function of buffer capacity **B** and initial marking **M**.

7.4   INSENSITIVITY

Insensitivity is the property that states that the stationary characteristic of the stochastic process underlying the queueing network depends on the service requirements only in terms of their averages. Product form queueing networks without blocking have been proved to be insensitive, i.e., the stationary joint queue length has been proved to depend on the service time distributions only in terms of their means.

Insensitivity can be extended to a certain class of queueing networks with finite capacity and blocking.

Referring to product form networks with finite capacity, Tables 5.1 and 5.2 and product form definitions show the cases where the stationary state distribution at arbitrary times depends on the distribution of the service time only in terms of the mean value (or the service rate μ_i).

Specifically, this insensitivity property holds for product form solutions PF1 which allow BCMP nodes, and for product form PFi, i=3,4,5 and 6 which allow A-type nodes, as defined in Section 5.1

Insensitivity for a two-node network with multiple class and RS blocking has been shown both for the joint stationary state distribution and for the call congestion of a job, i.e. the stationary probability that a job is blocked when requesting service at the next node.

Insensitivity of the joint queue length distribution holds for the central server and for reversible routing networks with A-type nodes and RS-RD and Stop blocking types.

7.5   MONOTONICITY

Some monotonicity properties of the network throughput hold for closed cyclic networks and for closed tandem networks with various blocking types, by considering increasing service rates or finite capacities or the overall network population.

For the special class of symmetrical networks introduced in Chapter 5, some equivalence results have been obtained for RS-RD blocking type in terms of throughput. Specifically, closed cyclic networks have the same throughput for N and MB-N customers both for exponential service times and phase-type distributions, i.e.,

$$X(N) = X(MB-N).$$

Moreover, such symmetry property can be related to network reversibility. This result is generalized to Fork/Join Queueing Networks with Blocking with more general service time distributions as presented in Section 7.3, Theorem 7.6.

Cyclic closed networks

This symmetry property holds for closed cyclic networks with BBS-SO blocking and PERT-type service time distribution and the throughput is maximized when the network population is $N = \lfloor B/2 \rfloor$ or $N = \lceil B/2 \rceil$.

Open tandem networks

One can prove that two tandem networks W and W^R with the same number of service centers and parameters, with BAS or BBS-SO blocking, but where the queues are the reverses of one another and full dual (see Section 7.3) have the same throughput. This is a special case of theorem 7.6 given in Section 7.3.

Let L_i and L_i^R denote the average queue length of service center i in network W and W^R, respectively. Then the following relation holds:

$$L_i^R = B - L_i.$$

For the special case of open tandem exponential networks with BBS-SO blocking and with $B_i = B_{M-i}$, $\mu_i = \mu_{M-i+1}$, $1 \leq i \leq M$, the following relation holds for the average queue length of node i, $1 \leq i \leq M$:

$$L_i = B_{M-i} - L_{M-i}$$

and if the number of nodes is even then the middle buffer is half full, on the average, regardless of the values of the other network parameters, i.e.:

$$L_j = B_j / 2, \quad \text{for } j = M / 2.$$

These results can be useful in checking the solution obtained by approximate numerical technique or simulation to analyze a tandem or cyclic queueing network with blocking.

Concavity
The throughput has been proved to be a concave function with respect to the buffer capacities.

Some monotonicity properties of the network throughput for increasing service rates or finite capacities or the overall network population have been proved for closed cyclic networks with BBS-SO blocking type. In other words, let $X(\mathbf{B})$ denote the network throughput as a function of the buffer capacities $\mathbf{B} = (B_1, ..., B_M)$, then the following relation holds between the throughput of the network with capacities $\mathbf{B}$ and capacities $\mathbf{B+k}$:

$$X(\mathbf{B+k}) \geq X(\mathbf{B}), \quad \text{for } k \geq 0.$$

This property holds for general service time distribution. In particular, this concavity property of network throughput holds for FJQNB with PERT type service time distribution, as stated by theorem 7.7 in Section 7.3. In these models the throughput increases as any buffer capacity increases.

Some comparison results between BAS and BBS blocking in terms of throughput can be obtained for this class of networks by considering various loading and unloading policies and characteristics of the fork and join node servers. The interested reader may refer to the references, see Section 7.6.

Concavity of the throughput may lead to the solution of optimization problems or to the definition and computation of throughput bounds.

Symmetrical networks
We shall now focus on the special class of closed symmetrical networks. In the following Section we present a methodology that allows to rank symmetrical networks with the same parameters (M, N, B, blocking type, μ) but different topology, according to their performance. We analyze the influence of topology on network performance. The choice of the system topology may presents performance advantages and disadvantages depending on the application field and the communication protocol used among the nodes. We focus on symmetrical queueing networks because such networks can be used to model systems with symmetrical interconnection topology, which often is a preferable choice for reasons of design simplicity and load balancing.

7.5.1 Performance comparison of blocking symmetrical networks

As we have seen in the previous chapters, in general the performance evaluation of networks with blocking is computational expensive. As a consequence, along with the search for efficient solution methodologies it is interesting to define methodologies that allow to predict and give some indications about the performance behavior of blocking networks without solving the models.

In this Section we investigates the possibility of extracting from the set of network parameters and in particular from the adjacency matrix an index that describes the network topology. Such index still retains all the relevant information for performance prediction and allows to compare the performance behavior of networks with different topologies. The computational cost of this index is significant lower than the computational cost of the network solution, but a conjecture states that by evaluating this index for some symmetrical networks we can establish an order among their relative performance. The conjecture is supported by a wide experimental analysis of several symmetrical networks.

Let us consider the class of symmetrical networks as defined in Chapter 5. We shall now define an index, called Cycle Frequency vector (CF-vector), which "counts" the number of cycles in the network and we shall see how the network performance depends also on this index.

In particular we consider closed homogeneous symmetrical networks with M exponential nodes, with single servers and constant service rates μ, node's buffer capacity B, a population of N jobs and blocking type $X \in \{RS-RD, BAS, BBS-O\}$. The scheduling algorithm at the queues are of abstract type, i.e., independent of the service demand.

Since we are interested in a comparison between the relative performance of different networks, in the following we consider the network effective utilization as performance index as a function of blocking type and topology. Indeed for symmetrical networks, utilization, throughput and mean response time provide the same information about the performance behaviour (see also Section 5.3.2).

The direct connection between nodes is specified by the (MxM) adjacency matrix $r=[r(i,j)]$, where $r(i,j)=1$ if there exists a link from i to j, $r(i,j)=0$ otherwise. Two nodes i and j such that $r(i,j)=1$ are called neighbors. The definition of symmetrical network given in Chapter 5 guarantees that rows 2,..., M of the adjacency matrix are circular permutations of the first row. Given the number of nodes M, $M \geq 3$, let r denote the outdegree of each node, $1 \leq r \leq M-1$. Then there are $u=\binom{M-2}{r-1}$ different adjacency matrices, i.e. topologies. The various topologies can systematically be derived as follows. Let $T_{r,t}$ denote the t-th topology of the network with degree r and $r_{r,t}$ and $r_{r,t}[1,*]$ the corresponding t-th adjacency matrix and its first row, respectively. Without loss of generality, we always choose the second node as the first neighbor of node 1. The remaining (r-1) nodes are selected in all possible ways over the remaining M-2 nodes. The first row of $r_{r,t}$ is $01c_1c_2 \ldots c_{M-2}$, where $c_1c_2 \ldots c_{M-2}$ is a sequence of binary digits where exactly r-1 of them are equal to 1; each other row of matrix $r_{r,t}$ is just a circular permutation of the first one. Hence the number of all network topologies is $u=\binom{M-2}{r-1}$. We assume a decreasing lexicographic order of the different topologies. For example, figure 7.7 illustrates the set of possible topologies for M=5 and r=2. For the sake of simplicity, the neighbors are depicted only for node 1.

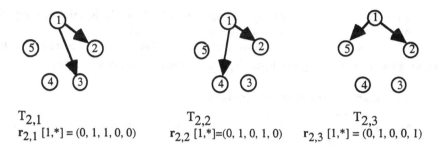

$T_{2,1}$ $T_{2,2}$ $T_{2,3}$

$r_{2,1} [1,*] = (0, 1, 1, 0, 0)$ $r_{2,2} [1,*]=(0, 1, 0, 1, 0)$ $r_{2,3} [1,*] = (0, 1, 0, 0, 1)$

Figure 7.7. Set of possible topologies for a 5-node-network with r=2. Links are depicted only for node 1

Let us define an index associated with a symmetrical network. This index consists of a vector whose components m_l are just the number of cycles of length l in the network, with $l=2, 3, ..., M$. We shall see that this index characterizes the topology of a symmetrical network with M nodes. In the following, the specification of the reference node is dropped from the notations since, in a symmetrical network, the number m_l is the same for each reference node.

The index associated with the network with topology $T_{r,t}$, r=1,..., M-1 and t=1,..., u, is called Cycle Frequency vector, denoted by $CF_{r,t}$, and consists of a vector whose k-th component is defined as

$$CF_{r,t} (k) = r^{k+1}_{r,t} (1,1) \qquad\qquad k=1, ... M-1.$$

where $r^{k+1}_{r,t}$ is the (k+1)-th power of the adjacency matrix $r_{r,t}$. The element $CF_{r,t} (k)$ represents the number of cycles with length (k+1) for each node of the network. This number is greater than or equal to the number of elementary cycles, since a cycle with length $l \leq N$ is not necessarily elementary, but all the elementary cycles are included in the CF-vector (for further details see the references in Section 7.6). Note that an *elementary cycle* is defined as one that does not contain the same node twice except for the ends.

Note that in the general case two symmetrical networks with the same number of nodes M, degree r, and different adjacency matrices have different CF-vectors, but in some cases can also have the same CF-vector.

Consider the lexicographic order on the set of CF-vectors, that is given two n-vectors **A** and **B** we define

$$A < B \text{ if } \exists j° \in \{1, ..., n\} : A(i) = B(i) \ 1 \leq i < j° \wedge A(j°) < B(j°).$$

Hence, we say that $CF_{r,t}$ follows $CF_{r',t'}$ if $CF_{r',t'} < CF_{r,t}$.

The relation " < " among CF-vectors induces an order among the corresponding symmetrical networks. Indeed, we say that a symmetrical network with topology $T_{r,t}$ is "more cyclic" than another, with the same M and topology $T_{r',t'}$, if $CF_{r',t'} < CF_{r,t}$. This order relation implies that a network is more cyclic than another one if it either has shorter cycles, or it has more cycles with length l, where l is the minimum length for which the two topologies have a different number of cycles.

In conclusion, the order among the $\mathbf{CF}_{r,t}$, defined by "<", corresponds to ranking the corresponding topologies $T_{r,t}$ according to increasing cyclicity.

Performance prediction via CF-vector

We shall now use the CF-vector to predict the performance of a symmetrical network.

The following theorem states that the performance of a symmetrical network with M nodes depends on the CF-vector. In other words the influence of the network topology on the system performance is completely captured by the CF-vector. For the proof see the reference in Section 7.6.

Theorem 7.8

The performance of a symmetrical network with M nodes and adjacency matrix $\mathbf{r}$, is completely determined by the following parameters:

$$(N, B, \text{blocking type}, \mu, \mathbf{CF})$$

where $\mathbf{CF}$ is the network cycle-frequency vector.

Corollary 7.2

Symmetrical networks with M nodes and the same parameters (N, B, blocking type, μ), different adjacency matrix but the same CF-vector show the same performance.

The proof immediately follows from Theorem 7.8.

The corollary states that identical CF-vectors imply the same performance for the corresponding networks. On the other hand, the following lemma can be used to prove that the opposite is not always true, since it shows that there exist symmetrical networks with same performance and different CF-vectors.

Lemma 7.1

Let R and R' be two symmetrical networks with M nodes and the same parameters (N, B, μ), RS-RD blocking, bidirectional links and different degrees r and r' ($1 \le r, r' \le M$). Then R and R' have the same performance but different CF-vectors.

Proof

The two networks have the same performance since bidirectional links lead to reversible routing which is a condition for product form solution to hold for RS-RD, as stated in Chapter 5. In this case, the calculation of the performance indices is based on probabilities that are independent of the degree and the routing matrix. On the other hand, it is easy to be convinced that networks with different degree r have in general different CF-vector. **QED**

Theorem 7.8 states dependence between the performance of a symmetrical network and its CF-vector but without any quantitative indication. If two networks have the same CF-vector, it is possible to conclude from the Corollary 7.2 that they have the same performance. On the other hand, if the two networks have different CF-vectors we cannot derive any performance relationship.

We shall now formulate a conjecture on the quantitative dependence between the performance of a symmetrical network and its CF-vector, supported by several numerical results. As stated above, we consider node effective utilization as the performance index.

To illustrate the conjecture we present some numerical results.

Example 7.4. Tables 7.5 and 7.6 show the utilization and the CF-vectors for the parameter set ($M=7$, $N=5$, $B=3$, $\mu=1$) and for the three considered blocking models. The population has been kept upper bounded so as to prevent deadlock situations as discussed in chapter 2, Section 2.2.

The tables have M-1 columns, for each degree $r=1, 2, ..., M-1$. The r-th column shows the results for each network topology of degree r $T_{r,t}$ for $t =1,..., u=\binom{M-2}{r-1}$

Each entry of Table 7.5 for column r and topology t gives the utilization for each blocking type, that is RS-RD, BAS and BBS-O blocking. Each entry of Table 7.6 gives the corresponding CF-vector.

Let us consider in particular topologies $T_{6,1}$, $T_{3,3}$, $T_{3,4}$, $T_{2,4}$, $T_{2,2}$ and $T_{1,1}$. By Table 7.6 we observe the following order of the CF-vectors:

$$CF_{6,1} > CF_{3,3} > CF_{3,4} > CF_{2,4} = CF_{2,2} > CF_{1,1}$$

while by Table 7.5 the comparison of their utilization when RS-RD blocking is used yields:

$$U^X_{6,1} > U^X_{3,3} > U^X_{3,4} > U^X_{2,4} = U^X_{2,2} > U^X_{1,1} \qquad X= \text{RS-RD}$$

On the contrary, for the same networks with BAS or BBS-O blocking the order is:

$$U^X_{6,1} < U^X_{3,3} < U^X_{3,4} < U^X_{2,4} = U^X_{2,2} < U^X_{1,1} \qquad X= \text{BAS, BBS-O}$$

Then we can argue a possible relation between network performance and its cyclicity characteristics that are captured by the CF-vector.

A wide sperimentation strongly confirms the possible relation described by Example 7.4, as expressed by the following conjecture.

Conjecture
The node effective utilization of a symmetrical network with parameters (M, N, B, μ) is monotonic increasing for RS-RD blocking and monotonic decreasing for BBS-O or BAS blocking with respect to the CF-vector. More formally, let $T_{r,t}$ and $T_{r',t'}$

be the topologies of two symmetrical networks with the same parameters (M, N, B, μ), then

$$CF_{r,t} \geq CF_{r',t'} \quad \text{if and only if} \quad U^{RS\text{-}RD}_{r,t} \geq U^{RS\text{-}RD}_{r',t'} \quad \text{or}$$
$$U^{X}_{r,t} \leq U^{X}_{r',t'} \quad X = BAS, BBS\text{-}O$$

Table 7.5. Node utilization for symmetrical networks with parameters (M=7, B=3, N=5, μ=1), for each degree and topology and RS-RD, BAS or BBS-O blocking

degree r	1	2	3	4	5	6
t=1	$T_{1,1}$	$T_{2,1}$	$T_{3,1}$	$T_{4,1}$	$T_{5,1}$	$T_{6,1}$
RS-RD	0.4527	0.4595	0.4614	0.4621	0.4623	0.4624
BAS	0.4544	0.4531	0.4526	0.4522	0.4521	0.4516
BBS-0	0.4527	0.4359	0.4205	0.4045	0.3963	0.3898
t=2		$T_{2,2}$	$T_{3,2}$	$T_{4,2}$	$T_{5,2}$	
RS-RD	——	0.4612	0.4613	0.4620	0.4623	——
BAS		0.4528	0.4526	0.4522	0.4521	
BBS-0		0.4293	0.4212	0.4073	0.3963	
t=3		$T_{2,3}$	$T_{3,3}$	$T_{4,3}$	$T_{5,3}$	
RS-RD	——	0.4595	0.4621	0.4620	0.4623	——
BAS		0.4531	0.4522	0.4522	0.4521	
BBS-0		0.4359	0.4096	0.4073	0.3963	
t=4		$T_{2,4}$	$T_{3,4}$	$T_{4,4}$	$T_{5,4}$	
RS-RD	——	0.4612	0.4619	0.4621	0.4623	——
BAS		0.4528	0.4523	0.4522	0.4521	
BBS-0		0.4293	0.4139	0.4045	0.3963	
t=5		$T_{2,5}$	$T_{3,5}$	$T_{4,5}$	$T_{5,5}$	
RS-RD	——	0.4624	0.4619	0.4621	0.4623	——
BAS		0.4520	0.4523	0.4522	0.4521	
BBS-0		0.4152	0.4139	0.4045	0.3963	
t=6			$T_{3,6}$	$T_{4,6}$		
RS-RD	——	——	0.4614	0.4624	——	——
BAS			0.4526	0.4519		
BBS-0			0.4205	0.3991		
t=7			$T_{3,7}$	$T_{4,7}$		
RS-RD	——	——	0.4621	0.4620	——	——
BAS			0.4522	0.4522		
BBS-0			0.4096	0.4073		
t=8			$T_{3,8}$	$T_{4,8}$		
RS-RD	——	——	0.4614	0.4624	——	——
BAS			0.4526	0.4518		
BBS-0			0.4526	0.3991		
t=9			$T_{3,9}$	$T_{4,9}$		
RS-RD	——	——	0.4621	0.4621	——	——
BAS			0.4522	0.4522		
BBS-0			0.4096	0.4045		
t=10			$T_{3,10}$	$T_{4,10}$		
RS-RD	——	——	0.4619	0.4620	——	——
BAS			0.4523	0.4522		
BBS-0			0.4139	0.4073		

This conjecture states that when the network works under BAS or BBS-O blocking, the greater the cyclicity the worse the network performance is. On the contrary, when the network works under RS-RD blocking, the greater the cyclicity the better the network performance is.

To illustrate the conjecture, let us consider the topologies $T_{2,1}$ and $T_{2,u}$ with u = M-2. It is easy to be convinced that $CF_{2,1} \leq CF_{2,M-2}$. Consider a node i and its two downstream nodes d_1 and d_2 as shown in figure 7.8.

For topology $T_{2,1}$ the downstream nodes are strictly dependent because they are

Table 7.6. CF-vectors for symmetrical networks with M=7 nodes and each degree and topology

degree r	1	2	3	4	5	6
t=1 $T_{r,1}$	$CF_{1,1}$ (0,0,0, 0,0,1)	$CF_{2,1}$ (0,0,4, 10,6,2)	$CF_{3,1}$ (0,6,16,20, 96,395)	$CF_{4,1}$ (2,12,30,150, 608,2258)	$CF_{5,1}$ (4,18,88,450, 2224,11177)	$CF_{6,1}$ (6,30,186, 1110,6666, 39990)
t=2 $T_{r,2}$	——	$CF_{2,2}$ (0,3,0, 5,15,2)	$CF_{3,2}$ (0,6,12,30, 108,318)	$CF_{4,2}$ (2,9,38,145, 587,2335)	$CF_{5,2}$ (4,18,88,450, 2224,11177)	——
t=3 $T_{r,3}$	——	$CF_{2,3}$ (0,0,4, 10,6,2)	$CF_{3,3}$ (2,3,10,45, 71,395)	$CF_{4,3}$ (2,9,38,145, 587,2335)	$CF_{5,3}$ (4,18,88,450, 2224,11177)	——
t=4 $T_{r,4}$	——	$CF_{2,4}$ (0,3,0, 5,15,2)	$CF_{3,4}$ (2,3,10,35, 101,318)	$CF_{4,4}$ (2,12,30,150, 608,2258)	$CF_{5,4}$ (4,18,88,450, 2224,11177)	——
t=5 $T_{r,5}$	——	$CF_{2,5}$ (2,0,6, 0,20,2)	$CF_{3,5}$ (2,3,10,35, 101,318)	$CF_{4,5}$ (2,12,30,150, 608,2258)	$CF_{5,5}$ (4,18,88,450, 2224,11177)	——
t=6 $T_{r,6}$	——	——	$CF_{3,6}$ (0,6,16,20, 96,395)	$CF_{4,6}$ (4,6,44,130, 622,2258)	——	——
t=7 $T_{r,7}$	——	——	$CF_{3,7}$ (2,3,10,45, 71,395)	$CF_{4,7}$ (2,9,38,145, 587,2335)	——	——
t=8 $T_{r,8}$	——	——	$CF_{3,8}$ (0,6,16,20, 96,395)	$CF_{4,8}$ (4,6,44,130, 622,2258)	——	——
t=9 $T_{r,9}$	——	——	$CF_{3,9}$ (2,3,10,45, 71,395)	$CF_{4,9}$ (2,12,30,150, 608,2258)	——	——
t=4 $T_{r,10}$	——	——	$CF_{3,10}$ (2,3,10,35, 101,318)	$CF_{4,10}$ (2,9,38,145, 587,2335)	——	——

Figure 7.8. Downstream nodes d_1 and d_2 of a node i in the symmetrical topologies $T_{2,1}$ and $T_{2,M-2}$ of networks with M nodes and degree r=2

directly connected, as a consequence they cannot send the workload to disjoint subsets of nodes in the network. On the contrary for topology $T_{2,M-2}$ nodes d_1 and d_2 are not directly connected and can send the workload to two disjoint subsets of nodes in the network.

These different results obtained for the various topologies can also be interpreted as the possibility of distributing the workload in distinct areas of the network. Hence the network topology affects the performance depending on the blocking type. For example if only node d_2 is full, that is $n_{d_2}=B$, and node i completes a service of a job destined to d_2, we make some intuitive observations on the network behaviour with different blocking types:

RS-RD blocking

The recycling mechanism can lead to a reduced congestion of node i by serving jobs destined to the non-full node d_1. For topology $T_{2,1}$ the workload continues to burden the full area because d_1 is an upstream node of the full node d_2, while for topology $T_{2,M-2}$, the recycling mechanism is really useful.

Hence, the cyclicity is a favorable characteristic.

BAS blocking

Upon completion of the service destined to d_2, node i is blocked. As soon as d_2 releases a job, the job waiting in node i immediately moves to node d_2. In case of very cyclic topologies this mechanism can degrade the performance. Indeed for $T_{2,M-2}$ topology if node d_2 completes a service destined to i, by effect of simultaneous transitions the population of the nodes i and d_2 does not change. Indeed by observing only the state components corresponding to these nodes, for $T_{2,M-2}$ a transition takes place from the state $S=(\ldots (n_i,1,\varnothing), \ldots (n_{d_2},0,[i]), \ldots)$ towards the state $S'=(\ldots (n_i,0,\varnothing), \ldots (n_{d_2},0,\varnothing), \ldots)$ with rate $p_{d_2i}\,\mu$. On the other hand, for topology $T_{2,1}$, for any destination of the job in service in d_2, the transition is from S towards $S^*=(\ldots (n_i-1,0,\varnothing), \ldots (n_{d_2},0,\varnothing), \ldots)$ with the same rate. In other words, the BAS type characteristic of continuing the service until the completion yields a real progress of the workload in the case of a less cyclic topology.

Hence, the cyclicity is an unfavorable characteristic.

BBS-O blocking

The saturation of node d_2 yields the blocking of its upstream nodes. For topology $T_{2,1}$ the blocking of the nodes i and d_1 can have a positive effect because it prevents the workload from going to the full area. On the contrary for topology $T_{2,M-2}$ the unconditional blocking can prevent the progress of workload towards network non-full areas.

Hence, the cyclicity is an unfavorable characteristic.

Computational complexity

In general, the performance evaluation of a blocking network requires the solution of the global balance equation system as discussed in Chapter 4 with a cost $O(|E|^3)$, where E is the state space and $|E|$ grows exponentially with the network parameters (M, N, B). The conjecture stated above for symmetrical networks allows to predict the average network performance behaviour of symmetrical networks as a function of the topology and of the blocking type with a significant lower cost, polynomial in the number of network nodes.

Indeed, let us consider the CF-vector computation. By definition it requires the computation of matrices r^k, k=2,..., M. Actually, only the first column of matrix r^k is needed. Let us denote r^k (*, 1) this column vector. Since r^k (*, 1) = r r^{k-1} (*, 1), the computation of r^k (*, 1) when r^{k-1} (*, 1) is known requires $O(M^2)$ operations in the worst case. Since this is repeated for k = 2 ,..., M, then the computation of the CF-vector needs $O(M^3)$ operations in the worst case and, of course, $O(M^3) << O(|E|^3)$.

7.6   BIBLIOGRAPHICAL NOTES

Equivalence properties between networks with and without blocking are given in Balsamo and Iazeolla (1983).

Equivalences between networks with different blocking types are given in Onvural and Perros (1986). A survey on equivalence properties of queueing networks with various blocking types and topologies with the references for each relationship can be found in Balsamo and De Nitto Personè (1994).

Equivalence between BBS and RS mechanism in two-node networks with multiclass BCMP type nodes and class independent capacities is proved in Onvural (1989). Equivalences between BBS and RS mechanism, BBS and BAS mechanism and BAS and the network without blocking, under different assumptions, for an exponential two-node network with Fork/Join centers are proved in De Nitto Personè and Grassi (1999). Note that this result is not reported in Table 7.2 and the interested reader can refer to this paper.

The special equivalence reported in Section 7.1.3, in the last row of Table 7.4 between an open network with Stop blocking and a closed network with an additional node and BBS-O blocking, is given in Balsamo and De Nitto Personè (1994), where the definition of Stop blocking functions a(n) and d(n) is introduced in Van Dijk (1991b). The relationship between any open queueing network with finite queues and Poisson arrivals and the correspondent closed network is discussed in Onvural (1990).

The equivalence results in terms of cycle time distribution for two node networks with BBS-SO or BAS blocking described in Section 7.2 and the proof of theorem 7.1 are given in Balsamo and Donatiello (1989a) and (1989b).

The class of Fork/Join Queueing Networks with Blocking and some equivalence properties for open exponential networks with BBS-SO blocking are introduced in Ammar and Gershwin (1989). Fork and join models for parallel processing systems can be found in Heidelberger and Trivedi (1982), Nelson, Towsley and Tantawi (1988) and Balsamo, Donatiello and Van Dijk (1998) and for manufacturing systems in Dallery and Gershwin (1992), Papadopoulus and Heavey (1996) and references therein.

Duality and reversibility in tandem networks was introduced in Gordon and Newell (1967), Yamazaki and Sakasegawa (1975), Muth (1979) and Melamed (1986).

Various properties including equivalence, monotonicity, symmetry and concavity of the system throughput in Fork/Join Queueing Networks with Blocking under general assumptions and with BBS-SO blocking are presented in Dallery, Liu and Towsley (1994). These properties are extended to BAS and BBS-O blocking and various loading and unloading disciplines for the same class of network models in Dallery, Liu and Towsley (1992). The proof of theorem 7.2 in Section 7.3 is due to Ammar and Gershwin (1989) for exponential networks and to Dallery, Liu and Towsley (1994) for more general service distributions. The equivalence between FJQNB and Strongly Connected Marked Graph and the proof of theorems 7.3 through 7.6 and Corollary 7.1 are given in this latter reference.

Insensitivity property of the stationary joint queue length distribution that depend on the service time distributions only in terms of their means in product form queueing networks without blocking are given in Baskett et al. (1975) and Chandy and Martin (1983). For networks with blocking insensitivity is discussed for two-node networks with multiple classes and RS blocking both for the joint stationary state distribution and for the call congestion of a job, in Van Dijk and Tijms (1986). Results on insensitivity of the joint queue length distribution for the central server and for reversible routing topology networks with A-type nodes and RS-RD and Stop blocking types are given in Yao and Buzacott (1985) and (1987), Akyildiz and Von Brand (1989a) and (1989b), Onvural (1989), Akyildiz and Van Dijk (1990) and Van Dijk (1991a). Insensitivity for Fork/Join Queueing Networks with Blocking is presented in Dallery, Liu and Towsley (1992) and (1994) for BAS and BBS-SO blocking.

Symmetry of the throughput for N and MB-N customers in closed cyclic networks with BBS-SO blocking type was proved for exponential service times in De Nitto Personè and Grillo (1987) and generalised to phase-type distributions in Dallery and Towsley (1991). The relationship between this symmetry property and reversibility is discussed in Dallery and Towsley (1991).

Some monotonicity properties of the network throughput for increasing service rates or finite capacities or the overall network population for closed cyclic networks with BBS-SO blocking type has been proved in Shantikumar and Yao (1989). Monotonicity of the throughput in feed-forward tree-like networks and assembly-type configurations modelled by open tandem networks with BAS blocking are given in Adan and Van Der Wal (1989). Concavity and monotonicity properties of the throughput in tandem networks with arbitrary service time discipline and BAS or BBS-SO blocking is given in Liu and Buzacott (1993). These results are generalized for the class of Fork/Join Queueing Networks with Blocking where monotonicity, symmetry and concavity properties of the network throughput are proved under general assumptions on the service time distribution in Dallery, Liu and Towsley (1994). Convexity properties of the network throughput in tandem open networks with generalized blocking are given in Cheng (1993). Some comparison results between different blocking mechanisms and loading/unloading discipline in Fork/Join Queueing Networks with Blocking are given in Dallery, Liu and Towsley (1992).

The methodology for symmetrical networks presented in Section 7.5.1 is given in De Nitto Personè (1994). Symmetrical queueing networks can be applied to model multicomputer interconnection networks and communication protocols and the impact of system topology design on system performance is discussed in Reed and Grunwald (1987). The methodology presented in Section 7.5.1 is based on some graph properties that can be found in Christofides (1975). The proof of theorem 7.8, corollary 7.2 and several numerical results are given in De Nitto Personè (1994).

REFERENCES

Adan, I., and J. Van Der Wal "Monotonicity of the throughput in single server production and assembly networks with respect to the buffer size" in *Queueing Networks with Blocking* (H.G. Perros and T. Altiok Eds.), Elsevier, 1989, 325-344.

Akyildiz, I.F., and N. Van Dijk "Exact Solution for Networks of Parallel Queues with Finite Buffers" in *Performance '90* (P.J. King 40, I. Mitrani and R.J. Pooley Eds.) North-Holland, 1990, 35-49.

Akyildiz, I.F., and H. Von Brand "Central Server Models with Multiple Job Classes, State Dependent Routing, and Rejection Blocking" IEEE Trans. on Softw. Eng., Vol. 15 (1989) 1305-1312.

Akyildiz, I.F., and H. Von Brand "Exact solutions for open, closed and mixed queueing networks with rejection blocking" J. Theor. Computer Science, 64 (1989) 203-219.

Ammar, M.H., and S.B. Gershwin "Equivalence Relations in Queueing Models of Fork/Join Networks with Blocking" Performance Evaluation, Vol. 10 (1989) 233-245.

Balsamo, S., and V. De Nitto Personè "A survey of Product-form Queueing Networks with Blocking and their Equivalences" Annals of Operations Research, Vol. 48 (1994) 31-61.

Balsamo, S., and L. Donatiello "On the Cycle Time Distribution in a Two-stage Queueing Network with Blocking", IEEE Transactions on Software Engineering, Vol. 13 (1989) 1206-1216.

Balsamo, S., and L. Donatiello "Two-stage Queueing Networks with Blocking: Cycle Time Distribution and Equivalence Properties", in *Modelling Techniques and Tools for Computer Performance Evaluation*, (R. Puigjaner, D. Potier Eds.), Plenum Press, 1989.

Balsamo, S., L. Donatiello and N. Van Dijk "Bounded performance analysis of parallel processing systems" IEEE Trans. on Par. and Distr. Systems, Vol. 9 (1998) 1041-1056.

Balsamo, S., and G. Iazeolla "Some Equivalence Properties for Queueing Networks with and without Blocking", in *Performance '83* (A.K. Agrawala, S.K. Tripathi Eds.) North Holland, 1983.

Baskett, F., K.M. Chandy, R.R. Muntz, and G. Palacios "Open, closed, and mixed networks of queues with different classes of customers" J. of ACM, Vol. 22 (1975) 248-260.

Chandy, K.M., A.J. Martin "A characterization of product-form queueing networks" J. ACM, Vol.30 (1983) 286-299.

Cheng, D.W. "Analysis of a tandem queue with state dependent general blocking: a GSMP perspective" Performance Evaluation, Vol. 17 (1993) 169-173.

Christofides, N. *Graph Theory - An Algorithmic Approach*. Academic Press, London, 1975.

Dallery, Y., and S.B. Gershwin "Manufacturing flow line systems: A review of models and analytical results" Queueing Systems, Vol. 12 (1992) 3-94.

Dallery, Y., Z. Liu, and D.F. Towsley "Equivalence, reversibility, symmetry and concavity properties in fork/join queueing networks with blocking" Techn, Report, MASI.90.32, Université Pierre et Marie Curie, France, June, 1990 and J. of the ACM, Vol. 41 (1994) 903-942.

Dallery, Y., Z. Liu, and D.F. Towsley "Properties of fork/join queueing networks with blocking under various blocking mechanisms" Techn, Report, MASI.92, Université Pierre et Marie Curie, France, April, 1992.

Dallery, Y., and D.F. Towsley "Symmetry property of the throughput in closed tandem queueing networks with finite buffers" Op. Res. Letters, Vol. 10 (1991) 541-547.

De Nitto Personè, V. "Topology related index for performance comparison of blocking symmetrical networks" European J. of Oper. Res., Vol. 78 (1994) 413-425.

De Nitto Personè, V., and D. Grillo "Managing Blocking in Finite Capacity Symmetrical Ring Networks" Third Int. Conf. on Data Comm. Systems and their Performance, Rio de Janeiro, Brazil, June 22-25 (1987) 225-240.

De Nitto Personè, V., and V. Grassi "An analytical model for a parallel fault-tolerant computing system" Performance Evaluation, Vol. 38 (1999) 201-218.

Gordon, W.J., and G.F. Newell "Cyclic queueing systems with restricted queues" Oper. Res., Vol. 15 (1967) 286-302.

Heidelberger, P., and S.K. Trivedi "Queueing Network models for parallel processing with synchronous tasks" IEEE Trans. on Comp., Vol. 31 (1982) 1099-1109.

Lam, S.S. "Queueing networks with capacity constraints" IBM J. Res. Develop., Vol. 21 (1977) 370-378.

Liu, X.G., and J.A. Buzacott "A decomposition related throughput property of tandem queueing networks with blocking" Queueing Systems, Vol. 13 (1993) 361-383.

Melamed, B. "A note on the reversibility and duality of some tandem blocking queueing systems" Manage. Sci., Vol. 32 (1986) 1648-1650.

Muth, E.J. "The reversibility property of production lines" Manage. Sci., Vol. 25 (1979) 152-158.

Nelson, R., D. Towsley and A. Tantawi "Performance analysis of parallel processing systems" IEEE Trans. on Softw. Eng., Vol. 14 (1988) 532-540.

Onvural, R.O. "A Note on the Product Form Solutions of Multiclass Closed Queueing Networks with Blocking" Performance Evaluation, Vol.10 (1989) 247-253.

Onvural R.O. "Survey of Closed Queueing Networks with Blocking" ACM Computing Surveys, Vol. 22 (1990) 83-121.

Onvural, R.O., and H.G. Perros "On equivalences of blocking mechanisms in queueing networks with blocking" Oper. Res. Letters, Vol. 5 (1986) 293-297.

Papadopoulus, H.T., and C. Heavey "Queueing Theory in manufacturing systems analysis and design: A classification of models for production and transfer lines" Europ. J. of Oper. Res., Vol. 92 (1996) 1-27.

Reed D.A., and D.C. Grunwald "The performance of Multicomputer Interconnection Networks" Computer, (1987) 63-73.

Shantikumar, G.J., and D.D. Yao "Monotonicity properties in cyclic networks with finite buffers" Proc. First Int. Workshop on Queueing Networks with Blocking, (H.G. Perros and T. Altiok Eds.) North Holland, 1989.

Van Dijk, N. "On 'stop = repeat' servicing for non-exponential queueing networks with blocking" J. Appl. Prob., Vol. 28 (1991) 159-173.

Van Dijk, N. "'Stop = recirculate' for exponential product form queueing networks with departure blocking" Oper. Res. Lett., Vol. 10 (1991) 343-351.

Van Dijk, N.M., and H.C. Tijms "Insensitivity in Two Node Blocking Models with Applications" in *Teletraffic Analysis and Computer Performance Evaluation* (Boxma, Cohen and Tijms Eds.), Elsevier Science Publishers, North Holland, 1986, 329-340.

Yamazaki, G., and H. Sakasegawa "Problems of duality in tandem queueing networks" Ann. Inst. Math. Stat., Vol. 27 (1975) 201-212.

Yao, D.D., and J.A. Buzacott "Modeling a class of state-dependent routing in flexible manufacturing systems" Annals of Oper. Res., Vol. 3 (1985) 153-167.

Yao, D.D., and J.A. Buzacott "Modeling a class of flexible manufacturing systems with reversible routing" Oper. Res., Vol. 35 (1987) 87-93.

8 BUFFER ALLOCATION IN QUEUEING NETWORKS WITH FINITE CAPACITIES

In this chapter we study the buffer allocation problem in queueing networks with finite capacities. We present some solution techniques proposed in the literature for queueing networks with arbitrary topology.

The performance of a system highly depends on its topology, routing in the network, and the capacities of its queues. Although a single optimization model may be formulated, for practical purposes, the problem is generally decomposed into three interrelated optimization problems:
   i) optimal topology problem
   ii) optimal routing problem
   iii) optimal resource allocation problem

Let us assume that the topology of the system and the routing in the network are given. The problem is then to determine the buffer capacities at service stations such that the network throughput is close to its maximum value. This problem has a long history of research and development. However, most of the studies reported in the literature consider tandem topologies of queueing networks in which service stations are connected in series, as illustrated in figure 8.1.

Figure 8.1. A five node tandem network

Tandem networks have been used to model a single line of multistage automated assembly lines or virtual paths in communications networks. But this model does not take into consideration the interactions between various tandem lines in the system. A heuristic has been proposed in literature to address the buffer allocation problem for tandem, merge and split topologies of automated assembly lines and a formula has been obtained for the buffer allocation in tandem configurations.

Let us consider arbitrary topologies of feed-forward open queueing networks. All service and interarrival times are assumed to be exponentially distributed. Due to the finite capacities of nodes, blocking occurs.

In this chapter, we present two approximation algorithms developed to allocate the buffer capacities at each node such that the network throughput is close to its optimum value. The buffer allocation problem is formulated in Section 8.1. In Section 8.2 we present an algorithm that combines dynamic programming with heuristics to solve open queueing networks with finite buffers. The dynamic programming approach, in this context, was first used in the optimal layout of a transfer line problem under the assumption of infinite node capacities at each node. Then it was extended to address buffer allocation problem in single transfer lines. In this respect, the first salgorithm presented here in Section 8.3 may be viewed as an extension of these approaches to arbitrary topologies of queueing networks. The second algorithm developed and presented in Section 8.4 is an extension of the dynamic programming approach that eliminates some of the inaccuracies introduced due to the use of dynamic programming in this context. We discuss the accuracy of the two algorithms in Section 8.5.

8.1   BUFFER ALLOCATION PROBLEM

Consider an open queueing network consisting of M nodes connected arbitrarily in a feed-forward manner. External arrivals to the network are assumed to arrive at node i in a Poisson manner with rate λ_i, $1 \leq i \leq M$. Arrival processes at nodes are assumed to be independent and identically distributed. A customer arriving at the network is assumed to be lost, if it finds the node full at that moment. The service time at node i is exponentially distributed with rate μ_i. Service times at each node are independent and identically distributed, and, are independent of the arrival processes. We assume that customers at each node are served in a FCFS manner. A customer completing its service at node i chooses its destination node j with probability p_{ij} or leaves the network with probability p_{i0}. We note that, due to the feed-forward connection of nodes, it is possible to number the nodes of the network such that $p_{ij}=0$ for $i \geq j$ and $1 \leq i,j \leq M$. A feed-forward five node open network is illustrated in figure 8.2.

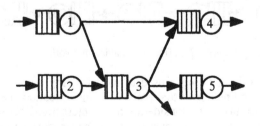

Figure 8.2. A feed-forward five node open queueing network

Due to limitations imposed on the buffer capacities, blocking occurs. The two blocking mechanisms used in this chapter are Blocking After Service (BAS) and Blocking Before Service (BBS) as defined in Chapter 2.

Simply defined, the buffer allocation problem is to allocate the buffer capacities between the service stations of a queueing network such that the throughput of the network is close to its optimum value, assuming that the network topology and routing probabilities are given. Let:

M	number of nodes
B	total number of buffer capacities to be allocated to M service stations
λ_i	arrival rate to node i, i=1,..., M
μ_i	service rate at service station i, i=1,..., M
p_{ij}	routing probability from node i to node j, i,j=1,..., M
C_i	the buffer capacity of node i, i=1,..., M; $0 \leq C_i \leq B$.

Note that the buffer capacity of a node does not include the server. That is $B_i = C_i + K_i$, where B_i is the overall node capacity an K_i is the number of servers.

$X(C_1, C_2,...,C_M)$ denote the throughput of the network with buffer capacities $(C_1, C_2,...,C_M)$.

Then, the optimal buffer allocation problem can be stated as follows:

$$\max X(C_1, C_2,...,C_M)$$
$$\text{subject to } \sum_{i=1}^{M} C_i = B$$
$$C_i > 0, C_i \text{ integer}$$

In this formulation of the buffer allocation problem, the cost related to providing buffers are determined solely by the total storage space and not by the number of allocations where buffers are provided. Various constraints on the buffer capacities may be included in a rather straightforward manner. The main difficulty in the above optimization problem is that the throughput of open queueing networks with finite buffers under the BAS and BBS blocking mechanisms do not have closed form solutions.

In the two algorithms presented in this chapter, a blocking network is solved a number of times with different buffer capacities. An algorithm to solve blocking networks is used to approximate the throughput of the blocking networks under consideration. Although there are more accurate algorithms reported in the literature, this particular algorithm is very easy to implement and produces fairly accurate results, particularly for tandem networks in general and arbitrary topologies of queueing networks under BBS blocking. However, any algorithm developed in the literature to solve open queueing networks with blocking can be used to replace this algorithm.

8.2 AN APPROXIMATION ALGORITHM FOR OPEN NETWORKS WITH BLOCKING

Consider an M node open network with exponentially distributed service and interarrival times under BBS-SO blocking. Let λ_i, μ_i, and p_{ij}, respectively denote the arrival and service rates at node i, and the probability that a customer upon completing its service at node i chooses node j as its destination node. The algorithm decomposes the network into individual nodes and analyzes each node in isolation. The service time of a node in isolation needs to be revised to accommodate the delays a node goes through due to blocking. Let f_{jk} and ϵ_j respectively denote the pseudo arrival rate from node j to node k, and probability that node j is full. Assuming for a moment that ϵ_j's are equal to the probabilities that node j blocks its arrivals and they are known a priori together with the pseudo arrival rates, each queue can then be solved as an M/M/1/C queue, where C is the capacity of the node excluding the server. Let ρ_j denote the pseudo utilization of node j in isolation. We note that, for the simplicity of notation, an empty sum in equation (8.1) equals to zero. Then:

$$\rho_j = \frac{\lambda_j + \sum_{i=1}^{j-1} f_{ij}}{\mu_j \Big/ \left(p_{j0} + \sum_{k=j+1}^{M} \frac{p_{jk}}{1-\epsilon_k} \right)} \qquad (8.1)$$

where the numerator of equation (8.1) is the total arrival rate and the denominator is the effective service rate at node j. The effective service process of a node in isolation is assumed to be exponentially distributed with its mean being approximated as the mean of the service process illustrated in figure 8.3.

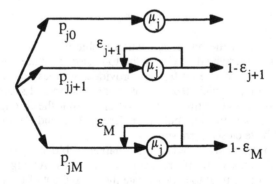

Figure 8.3. Effective Service Process at node j

In particular, if the customer at node j upon service completion departs from node j then the effective service process is the same as the original service process. However, if customer in service chooses node k as its destination node then server j is subject to blocking and the effective service process includes the blocking delays. Due to the memoryless property of exponential distribution, the remaining service time at node j when server j is blocked is the original service time. Assuming that ε_k is the blocking probability, blocked customer will receive geometrically distributed number of services. With this framework, the effective service process at node j when the customer in service chooses to go to node k is exponentially distributed with rate $\mu_j(1 - \varepsilon_k)$.

Using the steady state queue length distribution of an M/M/1/C queue, we have:

$$f_{jk} = \mu_j \frac{\rho_j - \rho_j^{C_j+2}}{1 - \rho_j^{C_j+2}} \frac{p_{jk}/1 - \varepsilon_k}{p_{j0} \sum_{l=j+1}^{M} \frac{p_{jl}}{1 - \varepsilon_l}} \qquad (8.2)$$

$$\varepsilon_j = \mu_j \frac{1 - \rho_j}{1 - \rho_j^{C_j+2}} \rho_j^{C_j+1} \quad j > 0 \text{ and } \varepsilon_0 = 0 \qquad (8.3)$$

The performance metrics of interest of a queueing network with blocking can be found iteratively using equations (8.1), (8.2), and (8.3). The throughput of the network with buffer capacities C_j, $X(C_1,C_2,...,C_M)$, is given as:

$$X(C_1,C_2,...,C_M) = \sum_{i=0}^{M} f_{i0} \qquad (8.4)$$

With BBS-SO blocking, the capacity of a node in isolation is the same as it is in the network. In case of BAS blocking, the buffer capacities of nodes are increased by the number of its upstream stations connected to the node, i.e. $C'_i = C_i + \sum_{k=1}^{M} 1(k,i)$, where $1(k,i)=1$ if $p_{ki}>0$ and 0 otherwise. This capacity adjustment, which is commonly used in various heuristics developed in the literature to analyze networks under BAS blocking, is required as a blocked customer in an upstream queue is in fact part of the queue of the full node that blocks its upstream nodes .

The accuracy of the two buffer allocation algorithms depends on the accuracy of the approximation algorithm used to solve the network. In particular, it is illustrated in Section 8.5 that when the approximate value of the network throughput is close to its exact value then the buffer allocation algorithms produce fairly accurate results, while the accuracy of the algorithms degrade as the difference between the exact and approximate values of the network throughput increases. We note that the approximation algorithm given in this section does not necessarily produce accurate results for arbitrary topologies of queueing networks under BAS blocking. However, the problem investigated in this chapter is the buffer allocation problem. Since the approximation algorithm used to obtain the network throughput is fairly accurate for BBS blocking and BAS blocking for tandem networks, the same algorithm is used for both blocking mechanisms in arbitrary topologies, mainly for the simplicity of the presentation.

8.3   DYNAMIC PROGRAMMING APPROACH

In the first approach, the optimal buffer allocation problem is formulated as a dynamic programming problem, and the approximation algorithm of Section 8.2 is used to solve the blocking network. The analysis of a node in isolation requires the arrival and effective service rates. Accordingly, a two-pass algorithm is developed: the forward problem where we assume that the blocking probabilities of nodes are known, and, the backward problem where it is assumed that the arrival rates are known. Then, the algorithm to obtain the buffer capacities iterates between the two passes until a convergence criterion is met. The notation used in the two buffer allocation algorithms are defined as follows:

- S_n: total buffer capacity allocated to nodes 1, 2,..., n
- $f_{ik}^t(n,S_n)$: pseudo-arrival rate from node i ($i \leq n$) to node k (k>n) given $C_1+C_2+...+C_n=S_n$ at the t-th iteration of the algorithm
- $\varepsilon_k^t(n,S_n)$: probability that node k (k>n) is full given $C_1+C_2+...+C_n=S_n$ at the t-th iteration of the algorithm
- $h\{\varepsilon_k^t(n,S_n)\}$: probability that node k is full conditioned upon a customer attempts to find out the number of customers in node k. Throughout the paper, it is assumed that $h\{\varepsilon_k^t(n,S_n)\}=\varepsilon_k^t(n,S_n)$.
- $g_{nk}(C_n,S_n)$: pseudo-arrival rate from node n to node k (k>n) given C_n and $C_1+C_2+...+C_n=S_n$.
- $\beta_n(C_n,S_{n-1})$: probability that node n is full given C_n and $C_1+C_2+...+C_{n-1}=S_{n-1}$.

8.3.1   Forward problem

The forward problem uses the pseudo-arrival rates to find the optimal buffer allocation that approximately maximizes the throughput of the network. We note that the parameters of the service process of a node can be easily calculated from the blocking probabilities $\varepsilon_k(n,S_n)$. Then, at the n-th stage of the forward problem, it is assumed that $\varepsilon_k(n,S_n)$'s, k=n+1,...,M, are known from the previous iteration of the backward problem. We then find the buffer allocation $C_1,C_2,...,C_n$ that maximizes the throughput across the cut between nodes i ($i \leq n$) and k (k>n) given $C_1+C_2+...+C_n= S_n$, using the pseudo arrival rates $f_{ik}(n-1, S_n-C_n)$ from node i ($i \leq n-1$) to node k ($k > n-1$) obtained at the (n-1)-st stage of the forward problem. Figure 8.4 illustrates the forward problem.

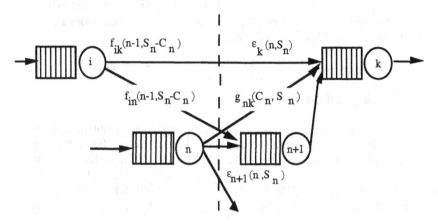

Figure 8.4. Forward Problem

The n-th stage at the t-th iteration of the algorithm is then given as follows: for n=1:

$$\max \sum_{k=n+1}^{M+1} [\lambda_k + g_{nk}(C_n,S_n)] [1 - h\{\varepsilon_k^{t-1}(n,S_n)\}] \quad \text{s.t. } 0 \leq C_n \leq S_n \quad (8.5.1)$$

for n=2,...,M-1

$$\max \sum_{k=n+1}^{M+1} [\lambda_k + \sum_{j=1}^{n-1} f_{jk}^t(n-1,S_n-C_n) + g_{nk}(C_n,S_n)] [1 - h\{\varepsilon_k^{t-1}(n,S_n)\}]$$

$$\text{s.t. } 0 \leq C_n \leq S_n \quad (8.5.2)$$

where k=M+1 denotes departures from the network. Furthermore, for simplicity of notation, we have $\lambda_{M+1}=0$, $h\{\varepsilon_{M+1}^{t-1}(n,S_n)\}=0$. Note that $g_{nk}(C_n,S_n)$ is obtained by the approximation algorithm given in Section 8.2 using $f_{ik}^t(n-1,S_n-C_n)$'s and $\varepsilon_k^{t-1}(n,S_n)$'s.

Once the optimal buffer capacity of node n, C_n^*, at the t-th iteration of the algorithm is found, the values of pseudo arrival rates and the buffer allocation are updated as follows:

$$f_{nk}^t(n,S_n)=g_{nk}(C_n^*,S_n), \quad\quad k=n+1,..., M+1$$
$$f_{ik}^t(n,S_n)=f_{ik}^t(n-1,S_n-C_n^*), \quad\quad i=1,..., n-1; \ k=n+1,..., M+1$$
$$C_n(n,S_n) = C_n^*$$

$$C_i(n,S_n) = C_i(n-1,S_n-C_n^*),\qquad i=1,...,\ n-1$$

where $C_1(n,S_n)$, $C_2(n,S_n)$,..., $C_n(n,S_n)$ is the buffer allocation which approximately maximizes the throughput across the cut between nodes i (i≤n) and node k (k>n) with $C_1+C_2+...+C_n=S_n$.

We now proceed with the description of the backward problem.

8.3.2 Backward problem

In the backward problem, the values of the pseudo-arrival rates obtained in the previous iteration of the forward problem are used to calculate the probability that node k is full, $\varepsilon_k(n,S_n)$. That is, at the n-th stage of the backward problem, it is assumed that the pseudo-arrival rates are known. Then, we find the buffer allocation C_n, C_{n+1},..., C_M that maximizes the throughput across the cut between nodes i (i<n) and k (k≤n) given $C_1+C_2+...+C_{n-1}=S_{n-1}$, using the values of $\varepsilon_k^t(n,S_{n-1}+C_n)$ obtained at the previous stage of the backward problem. Figure 8.5 illustrates the backward problem.

For each S_{n-1}, ($S_{n-1}=0,...,B$) the backward problem at the n-th stage of the t-th iteration can now be stated as follows:

for n=2,...,M:

$$\max\left[\lambda_n + \sum_{i=1}^{n-1} f_{in}^t(n-1,S_{n-1})\right]\left[1-\beta_n(C_n,S_{n-1})\right]$$

(8.6)

$$+ \sum_{k=n+1}^{M}[\lambda_n + \sum_{i=1}^{n-1} f_{ik}^t(n-1,S_{n-1})]\,[1-\varepsilon_k^t(n,S_{n-1}+C_n)]$$

$$\text{s.t. } 0\le C_n\le B\text{-}S_{n-1}$$

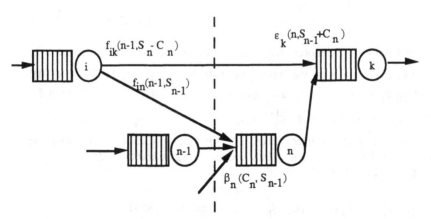

Figure 8.5. Backward Problem

Once the optimal buffer capacity of node n, C^*_n, at the t-th iteration of the algorithm is found, the values of the probability that node k is full ($k\leq n$) and the buffer allocation are updated as follows:

$$\varepsilon^t_n (n\text{-}1, S_{n\text{-}1}) = \beta_n (C^*_n, S_{n\text{-}1})$$

$$\varepsilon^t_k (n\text{-}1, S_{n\text{-}t}) = \varepsilon^t_k (n, S_{n\text{-}1} + C^*_n), \quad k = n+1, ..., M$$

$$C_n(n\text{-}1, S_{n\text{-}1}) = C^*_n$$

$$C_k(n\text{-}1, S_{n\text{-}t}) = C_k(n, S_{n\text{-}1} + C^*_n), \quad k = n+1, ..., M$$

where $C_n(n\text{-}1, S_{n\text{-}1}), C_{n+1}(n\text{-}1, S_{n\text{-}1}), ..., C_M(n\text{-}1, S_{n\text{-}1})$ is the buffer allocation that maximizes the throughput across the cut between node i ($i<n$) and node k ($k\leq n$) with $C_1+C_2+...+C_{n\text{-}1} = S_{n\text{-}1}$.

8.3.3 Algorithm

We note that, at the n-th stage of the forward and the backward problems, we maximize the total throughput on the cut that separates nodes 1 to n and n+1 to N. The forward problem requires the knowledge of the blocking probabilities while the arrival rates are assumed to be known in the backward problem. Since these parameters are not known in advance, the algorithm to obtain the buffer capacities that maximizes the network throughput iterates between the forward and backward problems until a convergence criteria is met. In particular, we choose to stop if the difference of the throughput values obtained in the forward and backward problems is negligible ($<10^{-5}$), or the two problems give the same allocation vector. The algorithm can now be summarized as illustrated in Table 8.1.

8.4 AN EXTENSION

At the t-th iteration of the dynamic programming approach for the n-th node, the blocking probabilities used in the forward problem are obtained from the previous iteration of the backward problem. These blocking probabilities are calculated assuming that the buffer capacities of nodes n+1 to M are allocated such that the throughput across the cut between two nodes i and k, $i\leq n$ and $k>n$, is maximized. However, these buffer capacities may not be optimal in the current iteration of the forward problem as the pseudo arrival rates in the two consecutive iterations of the algorithm may be quite different from each other. A similar problem may occur in the backward problem as the pseudo arrival rates used in the current iteration are the values obtained during the last forward iteration. Even though the algorithm iterates until a convergence criterion is met, the point of convergence, due to the above reasoning, may be different than the optimal one. In the following approach, the total throughput of the network as opposed to maximizing the total throughput over a cut is maximized.

Table 8.1. Algorithm 1

Algorithm 1:

begin
 for n =1 to M-1 do (* initialize the blocking probabilities *)
  for S_n =0 to B do
   for k =2 to M+1 do
   $\varepsilon^0_k(n,S_n)=0$;
 t =1;   (* iteration index is set to 1 *)
 A: (* update the values of buffer capacities and pseudo arrival rates *)
   for S_1 =0 to B do
    begin
     $C^t_1 = S_1$;
      for k =2 to M+1 do
           calculate $f^t_{1k}(1,S_1)$;
   end;
(* forward problem *)
for n =2 to M do
 for S_n =0 to B do
  begin
   solve the forward problem;
   for i =1 to n do
   for k =2 to M+1 do
     begin calculate new $f^t_{ik}(1,S_n)$; calculate new $C^t_i(n,S_n)$; end;
  end;
calculate the total throughput of the network;
if convergence is achieved then STOP;
(* backward problem *)
for S_{M-1} =0 to B do
 begin $C^t_M(M-1,S_{M-1}) = B - S_{M-1}$; calculate new $\varepsilon^t_M(M-1,S_{M-1})=0$;
 end;
for n =M-1 downto 1 do
 for S_{n-1} =0 to B do
  begin
   solve the backward problem;
   for k =n to M do
    begin calculate new $\varepsilon^t_k(n-1,S_{n-1})$; calculate new $C^t_k(n-1,S_{n-1})$;
   end;
  end;
 t =t+1;
Go to A;
end;

Similar to the dynamic programming approach, this algorithm iterates between the forward and the backward algorithms until a convergence criterion is met.

8.4.1 Forward problem

In the n-th stage of the forward problem, we obtain the optimal buffer allocation of the capacity S_n to nodes 1 to n. To do so, we search for the buffer capacity of node n, C_n, such that the throughput of the network is maximized subject to $0 \leq C_n \leq S_n$, using the optimal buffer allocation to all nodes other than node n. In particular, at the n-th stage, we use the optimal buffer allocation to nodes 1 to n-1 which was obtained at the (n-1)-st stage of the forward problem and the optimal buffer allocation (of the capacity $B-S_n$) to nodes n+1 to M which was obtained at the previous iteration of the backward problem. Then, the t-th iteration of the forward problem is given as follows:

$$\max X\{ C_1^t(n-1,S_n-C_n),..., C_{n-1}^t(n-1,S_n-C_n),C_n, C_{n+1}^{t-1}(n,S_n),..., C_M^{t-1}(n,S_n)\}$$
$$\text{s.t. } 0 \leq C_n \leq S_n \tag{8.7}$$

Then, for optimal $C^*{}_n$, we have:

$$C_n^t(n, S_n) = C^*{}_n$$
$$C_i^t(n, S_n) = C_i^t(n-1, S_n-C^*{}_n), i=1,...,n-1$$

where $C_i^t(n, S_n)$, i=1,...,n, is the buffer allocation that maximizes the total throughput of the network given the buffer capacities of nodes n+1 to M.

Similarly, the backward problem can be stated as follows.

8.4.2 Backward problem

In the n-th stage of the backward problem, we obtain the optimal buffer allocation of the capacity $B-S_{n-1}$ to nodes n to M. Then, the t-th iteration of the backward problem is given as follows:

$$\max \{X\{ C_1^t(n-1,S_{n-1}),..., C_{n-1}^t(n-1,S_{n-1}),C_n, C_{n-1}^t(n,S_{n-1}+C_n),...,$$
$$C_M^{t-1}(n,S_{n-1}+C_n)\}$$
$$\text{such that } 0 \leq C_n \leq B-Sn-1 \tag{8.8}$$

Then, for optimal $C^*{}_n$, we have:

$$C_n^t(n-1, S_{n-1}) = C^*{}_n$$
$$C_i^t(n-1, S_{n-1}) = C_i^t(n-1, S_{n-1}+C^*{}_n), i=n+1,...,M$$

where $C_i^t(n-1, S_{n-1})$, i=n,..., M, is the buffer allocation that maximizes the total throughput of the network given the buffer capacities of nodes 1 to n-1.

8.4.3 Algorithm

The algorithm iterates between the forward and the backward problems until both algorithms produce an identical set of buffer capacities. Table 8.2 shows

Table 8.2. Algorithm 2

Algorithm 2:

begin
  for n =1 to M-1 do
    for S_n =0 to B do
      for k =n+1 to M+1 do

      Allocate the total capacity to $C_k^0 (n, S_n)$ as equally as possible;

t =1;   (* iteration index is set to 1 *)
A: (* update the values of buffer capacities and pseudo arrival rates *)
    for S_t =0 to B do

$$C_1^t(1, S_1) = S_1;$$

(* forward problem *)
for n =2 to M do
  for S_n =0 to B do
    begin
      solve the forward problem;
      for i =1 to n do

      calculate new $C_i^t(n, S_n)$;

    end;
if convergence is achieved then STOP;
(* backward problem *)
for S_{M-1} =0 to B do

    $C_M^t (M-1, S_{M-1}) = B - S_{M-1};$

for n =M-1 downto 1 do
  for S_{n-1} =0 to B do
    begin
    solve the backward problem;
    for k =n to M do

      calculate new $C_k^t (n-1, S_{n-1});$

    end;
t =t+1;
Go to A;
end;

this second algorithm.

8.5   VALIDATION

The accuracy of the two buffer allocation algorithms presented in Sections 8.3 and 8.4 has been investigated for two different topologies, each with five nodes: tandem networks (figure 8.1), and arbitrary topologies (figure 8.2).

There are two main sources of inaccuracies in the above approaches: approximate solution of queueing networks with blocking, and, by the fact that the approaches themselves are approximations. We first investigate the accuracy of the two approaches with respect to the approximation algorithm used to solve blocking networks. In particular, let us for a moment assume that the approximation algorithm produces the exact throughput figures. Given the parameters, the topology of the network, and the total number of buffers B, the approximation algorithm is executed to obtain the throughput of the network for all possible values of buffer capacities of nodes, C_i, such that $\sum_{i=1}^{M} C_i = B$. At the end of this exhaustive search, let $C^*(i)$ be the set of buffer capacities of nodes 1 to M that produce the i-th maximum throughput. It is noted that the ordering is determined using the approximation algorithm. Furthermore, let HM and DP be the optimum set of buffer capacities of nodes 1 to M obtained using the Heuristic Method and Dynamic Programming, the two algorithms presented in Sections 8.3 and 8.4, respectively. In all examples, it was observed that $HM=C^*(1)$. That is, the heuristic method is always observed to produce the best buffer allocation based on the exhaustive search performed using the approximation algorithm. The dynamic programming approach, on the other hand, appears to produce relatively inaccurate results compared to the heuristic method. That is, it is observed that the buffer allocation produced using the DP algorithm is not always equal to $C^*(1)$. Nevertheless, the difference between the throughput values of the network with buffer capacities obtained using the two methods is observed to be insignificant in all examples we ran.

Tables 8.3 and 8.4 below present the parameters of the networks under consideration, the allocation of the buffer capacities obtained with the DP and HM methods, and how the results obtained with the dynamic programming approach (DP) compares to the best solution obtained with the exhaustive search (using the approximation algorithm) in the column #Best.

In tables 8.5 and 8.6, the throughput values (X) obtained by the two algorithms are compared to the throughput of the optimal buffer allocation. The optimal buffer allocation is determined through at least 30 configurations with buffer capacities around the buffer allocation obtained by the two algorithms. The RE% in these tables denotes the relative error percentages calculated by multiplying the absolute value of the difference between the optimal and approximate throughput by 100 and dividing by the value of the optimal throughput.

Table 8.3. Buffer Allocation-Tandem Network with M=5, B=10, λ_1=1

	Service Rates and Blocking Types	Buffer Allocation		
		DP	# Best	HM
T1	μ=(1.5,1.5,1.5,1.5,1.5), BAS: BBS:	(4,2,2,1,1) (3,2,2,2,1)	1 1	(4,2,2,1,1) (3,2,2,2,1)
T2	μ=(2.5,1.5,2.5,1.5,2.5), BAS: BBS:	(6,2,1,1,0) (4,2,2,1,1)	1 1	(6,2,1,1,0) (4,2,2,1,1)
T3	μ=(2.4,2.1,1.8,1.5,1.2), BAS: BBS:	(5,1,1,1,2) (3,1,2,2,2)	3 4	(4,1,1,2,2) (3,1,1,2,3)

Table 8.4. Buffer Allocation-Arbitrary Topology of figure 8.2 with M=5, λ_1=1.05, λ_2=0.45, p_{13}=0.5, p_{14}=0.5, p_{34}=0.4, p_{35}=0.3, p_{30}=0.3

	Service Rates and Blocking Types	Buffer Allocation		
		DP	# Best	HM(Best)
ARB 1 B = 10	μ=(1.3,0.55,1.35,1.1,0.45), BAS: BBS:	(6,2,1,1,0) (3,2,3,2,0)	2 5	(5,3,1,1,0) (4,1,2,2,1)
ARB 2 B = 15	μ=(1.3,0.55,1.35,1.1,0.45), BAS: BBS:	(8,3,2,2,0) (6,2,3,3,1)	3 2	(7,4,2,2,0) (5,3,3,3,1)
ARB 3 B = 20	μ=(1.3,0.55,1.35,1.1,0.45), BAS: BBS:	(9,4,3,3,1) (8,3,4,4,1)	14 3	(9,5,3,3,0) (7,4,4,4,1)
ARB 4 B = 10	μ=(1.3,0.5,1.35,1.1,0.35), BAS: BBS:	(5,2,2,1,0) (4,1,3,2,0)	1 9	(5,2,2,1,0) (4,1,2,2,1)
ARB 5 B = 15	μ=(1.3,0.5,1.35,1.1,0.35), BAS: BBS:	(7,3,3,2,0) (5,2,4,3,1)	6 1	(7,3,2,2,1) (5,2,4,3,1)
ARB 6 B = 20	μ=(1.3,0.5,1.35,1.1,0.35), BAS: BBS:	(9,3,4,3,1) (7,3,5,4,1)	44 6	(8,5,3,3,1) (7,3,4,4,2)
ARB 7 B = 10	μ=(1.3,0.55,1.35,1.1,0.3), BAS: BBS:	(5,2,2,1,0) (4,1,2,2,1)	2 1	(5,2,1,1,1) (4,1,2,2,1)
ARB 8 B = 15	μ=(1.3,0.55,1.35,1.1,0.3), BAS: BBS:	(7,3,3,2,0) (5,2,4,3,1)	58 2	(7,3,2,2,1) (5,2,3,3,2)
ARB 9 B = 20	μ=(1.3,0.55,1.35,1.1,0.3), BAS: BBS:	(7,3,4,4,2) (7,3,5,4,1)	104 58	(8,5,3,3,2) (7,3,4,4,2)

Table 8.5 illustrates a sample set of examples on the typical accuracy of the two approaches in tandem networks. In most cases, the buffer allocation determined by the two algorithms is observed to be the same as the optimal buffer allocation. In cases where the approximation algorithms do not produce the optimal allocation, the difference between the throughput of the tandem network with buffer capacities given by the algorithms and optimal allocation is fairly small as illustrated by the relative error percentages of less than 1%

Table 8.5. Throughput Comparison-Tandem Networks

	Bl.	Dynamic Programming		Heuristic Method		Optimal	
Ex.	Mec.	X	RE%	X	RE%	X	Buffer Alloc.
T1	BAS	0.88549	0	0.88549	0	0.88549	(4,2,2,1,1)
	BBS	0.82546	0	0.82546	0	0.82546	(3,2,2,2,1)
T2	BAS	0.96027	0.16%	0.96027	0.16%	0.96181	(6,1,2,1,0)
	BBS	0.92567	0.21%	0.92567	0.21%	0.92766	(3,2,2,2,1)
T3	BAS	0.92241	0.56%	0.92756	0	0.92756	(4,1,1,2,2)
	BBS	0.87737	0.58%	0.88249	0	0.88249	(3,1,1,2,3)

Table 8.6. Throughput Comparison-Arbitrary Topology

	Bl.	Dynamic Programming		Heuristic Method		Optimal	
Ex.	Mec.	X	RE%	X	RE%	X	Buffer Alloc.
ARB 1	BAS	1.07538	2.4%	1.07462	2.4%	1.10166	(5,2,1,2,0)
	BBS	1.20268	0	1.19485	0.65%	1.20268	(3,2,3,2,0)
ARB 2	BAS	1.15976	1.4%	1.15847	1.5%	1.17668	(7,3,3,2,0)
	BBS	1.29915	0.17%	1.30140	0	1.30140	(5,3,3,3,1)
ARB 3	BAS	1.28310	0	1.26514	1.4%	1.28310	(9,4,3,3,1)
	BBS	1.36289	0.2%	1.36560	0	1.36560	(7,4,4,4,1)
ARB 4	BAS	1.02751	0	1.02751	0	1.02751	(5,2,2,1,0)
	BBS	1.19676	0	1.18938	0.62%	1.19676	(4,1,3,2,0)
ARB 5	BAS	1.09166	6.6%	1.15701	1%	1.16880	(6,3,3,2,1)
	BBS	1.29229	0.23%	1.29229	0.23%	1.29524	(5,3,3,3,1)
ARB 6	BAS	1.22124	0	1.20802	1.1%	1.22124	(9,3,4,3,1)
	BBS	1.35737	0.03%	1.35421	0.26%	1.35775	(7,3,5,4,1)
ARB 7	BAS	0.97126	7.2%	1.02997	1.6%	1.04712	(4,2,1,2,1)
	BBS	1.18522	0	1.18522	0	1.18522	(4,1,2,2,1)
ARB 8	BAS	1.02541	8.3%	1.09782	1.8%	1.11817	(7,3,2,1,2)
	BBS	1.28837	0.04%	1.28098	0.62%	1.28888	(6,2,3,3,1)
ARB 9	BAS	1.20688	0	1.17285	2.8%	1.20688	(7,3,4,4,2)
	BBS	1.35360	0	1.35021	0.25%	1.35360	(7,3,5,4,1)

given in table 8.5. Finally, it is noted that the approximation algorithm used to obtain the throughput is observed to produce fairly accurate results in tandem networks under both blocking mechanisms.

In case of arbitrary topologies, the approximation algorithm used to solve the blocking network produces fairly accurate results under BBS blocking with relative error percentages of less than 4%.

Similarly, the two algorithms developed to determine the buffer allocation produces fairly accurate values. In particular, the two algorithms produce the optimal buffer allocation in a number of cases. Furthermore, the relative error percentages are observed to be less than 2% when sub-optimal values are produced.

However, the approximation algorithm used to solve blocking networks is not quite as accurate with BAS blocking. With relative error percentages of up to 30% in the throughput values, the accuracy of the dynamic programming approach degrades as illustrated in examples ARB-5, ARB-7, and ARB-8. However, in cases where the approximation algorithm produces fairly accurate results, the dynamic programming approach produce fairly accurate allocations. Unlike the dynamic programming approach, the accuracy of the heuristic method appears to be less affected by the accuracy of the approximation algorithm. In particular, the relative error percentages of the heuristic method is much less than they are in the dynamic programming approach in cases where the accuracy of the approximation algorithm is not good. On the other hand, dynamic programming approach produces relatively more accurate results than the heuristic method in cases where the approximation algorithm produces accurate solution to the blocking network.

8.6   BIBLIOGRAPHICAL NOTES

The buffer allocation problem in queueuing networks is formulated as single optimization model in Gavish (1986) and Gavish and Neuman (1986) and decomposed into three interrelated optimization problems in MacGregor et alt. (1988).

The tandem network to model single line of multistage automated assembly lines or virtual paths in communications networks is presented in Sevastyonov (1962), Buzacott (1967), (1968), (1971), Sheskin (1976), Soyster et al. (1979), Hillier and Boling (1966), (1967), (1979), Hillier and So (1989), Altiok and Stidham (1983), Gershwin (1984), (1987) and references there in.

The buffer allocation problem tandem, merge and split topologies of automated assembly lines was solved with a heuristic in MacGregor et alt. (1988), and with a formula for tandem configurations in Buzacott and Shanthikumar (1992).

We have presented two algorithms to obtain the buffer capacities of open queueing networks with arbitrary topologies, exponentially distributed service and interarrival times. The algorithm presented in Section 8.3 combines dynamic programming with heuristics. In this context the dynamic programming approach was first used by Kubat and Sumita (1985) in the optimal layout of a transfer line problem under the assumption of infinite node capacities at each node. Yamashita and Suzuki (1987) and Jafari and Shanthikumar (1989) extended the dynamic programming approach to address buffer allocation problem in single transfer lines.

General service and arrival processes can be easily incorporated into the two proposed approaches. In particular, these parameters only affect the methodology used to solve the blocking networks. Accordingly, by utilizing the algorithms developed in the literature to analyze more general queueing networks with blocking, the two approaches can be used without any changes to determine the buffer capacities.

REFERENCES

Altiok, T., and S.S. Stidham "The Allocation of Interstage Buffer Capacities in Production Lines" IIE Trans., Vol. 15 (1983) 292-299.

Buzacott, J. "Automatic Transfer Lines with Buffer Stocks" Int. J. Prod. Res., Vol. 5 (1967) 183-200.

Buzacott, J. "Prediction of the Efficiency of Production Systems without Internal Storage" Int. J. Prod. Res., Vol. 6 (1968) 173-188.

Buzacott, J. "The Role of Inventory Banks in Flow Line Production Systems" Int. J. Prod. Res., Vol. 9 (1971) 425-436.

Buzacott, J., and J.G. Shanthikumar "Design of Manufacturing Systems using Queueing Models" Queueing Systems: Theory and Applications, (1992).

Gavish, B. "A General Model for Topological Design of Computer Networks" in Proc. Globecom'86, 1986.

Gavish, B., and I. Neuman "Capacity and Flow Assignments in Large Computer Networks" in Proc. IEEE-Infocom'86, 1986, 275-284.

Gershwin, S. B. "An Efficient Decomposition Method for the Approximate Evaluation of Production Lines with Finite Storage Space" in Proc. Sixth INRIA Conf. Proc., Nice, France, 1984.

Gershwin, S. B. "An Efficient Decomposition Method for the Approximate Evaluation of Tandem Queues with Finite Storage Space and Blocking" Oper. Res., Vol. 35 (1987) 291-305.

Hillier, F.S., and W. Boling "The Effect of Some Design Factors on the Efficiency of Production Lines with Variable Operation Times" J. Ind. Eng., Vol. 7 (1966) 651-658.

Hillier, F.S., and W. Boling "Finite Queues in Series Exponential or Erlang Service Times-A Numerical Approach," Oper. Res., Vol. 15 (1967) 286-303.

Hillier, F.S., and W. Boling "On the Optimal Allocation of Work in Symmetrically Unbalanced Production Line Systems with Variable Operation Times" Mgmt. Sci., Vol. 25 (1979) 721-728.

Hillier, F.S., and K.C. So "The assignment of extra servers to stations in tandem queueing systems with small or no buffers" Performance Evaluation, Vol. 10 (1989) 213-231.

Jafari, M. A., and J.G. Shanthikumar "Determination of Optimal Buffer Storage Capacities and Optimal Allocation in Multistage Automatic Transfer Lines" IIE Trans., Vol. 21 (1989) 130-135.

Kubat, P., and U. Sumita "Buffers and Backup Machines in Automatic Transfer Lines" Int. J. Prod. Res., Vol. 23 (1985) 1259-1280.

MacGregor Smith, J., and S. Daskalaki "Buffer Space Allocation in Automated Assembly Lines" Operations Research, Vol. 36 (1988) 343-358.

Sevastyonov, B.A. "Influence of Storage Bin Capacity on the Average, Standstill Time of Production Line" Theory of Probability and Its Applications, Vol. 7 (1962) 429-438.

Sheskin, T. J. "Allocation of Interstage Storage Along an Automated Production Line" AIIE Trans., Vol. 8 (1976) 146-152.

Soyster, A.L., J. W. Schmidt and M.W. Rohrer, "Allocation of Buffer Capacities for a Class of Fixed Cycle Production Lines" AIIE Trans., Vol. 11 (1979) 140-146.

Yamashita, H., and S. Suzuki "An Approximate Solution Method for Optimal Buffer Allocation in Serial n-stage Automatic Production Lines" Trans. Japan. Soc. Mech. Eng., Vol. 53-C (1987) 807-814, (in Japanese).

REFERENCES

Adan, I., and J. Van Der Wal "Monotonicity of the throughput in single server production and assembly networks with respect to the buffer size" in *Queueing Networks with Blocking* (H.G. Perros and T. Altiok Eds.), Elsevier, 1989, 325-344.

Akyildiz, I.F. "Exact Product Form Solutions for Queueing Networks with Blocking" IEEE Trans. on Computers, Vol. 1 (1987) 121-126.

Akyildiz, I.F. "General Closed Queueing Networks with Blocking" in *Performance '87*, Courtois and Latouche (Eds), 282-303, Elsevier Science Publishers (North Holland), Amsterdam, 1988.

Akyildiz, I.F. "On the Exact and Approximate Throughput Analysis of Closed Queueing Networks with Blocking" IEEE Trans. on Software Engineering, Vol. 14 (1988), 62-71.

Akyildiz, I.F. "Mean Value Analysis for Blocking Queueing Networks" IEEE Trans. on Software Engineering, Vol. 14 (1988) 418-429.

Akyildiz, I.F. "Product Form Approximations for Queueing Networks with Multiple Servers and Blocking" IEEE Trans. Computers, Vol. 38 (1989) 99-115.

Akyildiz, I.F. "Analysis of Queueing Networks with Rejection Blocking" in Proc. First International Workshop on Queueing Networks with Blocking, Perros and Altiok (Eds), Elsevier Science Publishers, (North Holland), Amsterdam, 1989.

Akyildiz, I.F., and S. Kundu "Deadlock Free Buffer Allocation in Closed Queueing Networks" Queueing Systems Journal, 4 (1989) 47-56.

Akyildiz, I.F., and J. Liebeherr "Application of Norton's Theorem on Queueing Networks with Finite Capacities" in Proc. INFOCOM 89, 914-923, 1989.

Akyildiz, I.F., and H.G. Perros Special Issue on Queueing Networks with Finite Capacity Queues, Performance Evaluation, Vol. 10, 3 (1989).

Akyildiz, I.F., and N. Van Dijk "Exact Solution for Networks of Parallel Queues with Finite Buffers" in: Proc. *Performance '90* (P.J.B. King, I. Mitrani and R.J. Pooley Eds.) North-Holland (1990) 35-49.

Akyildiz, I.F., and H. Von Brand "Duality in Open and Closed Markovian Queueing Networks with Rejection Blocking" Tech. Rep., CS-87-011, (1987), Louisiana State University.

Akyildiz, I.F., and H. Von Brand "Exact solutions for open, closed and mixed queueing networks with rejection blocking" J. Theor. Computer Science, 64 (1989) 203-219.

Akyildiz, I.F., and H. Von Brand "Computation of Performance Measures for Open, Closed and Mixed Networks with Rejection Blocking" Acta Informatica (1989).

Akyildiz, I.F., and H. Von Brand "Central Server Models with Multiple Job Classes, State Dependent Routing, and Rejection Blocking" IEEE Trans. on Softw. Eng., Vol. 15 (1989) 1305-1312.

Allen, A.O. *Probability, Statistics and Queueing Theory with Computer Science Applications.* Academic Press, New York, 1990.

Altiok, T., and H.G. Perros "Open networks of queues with blocking: split and merge configurations" IEE Trans. 9 (1986) 251-261.

Altiok, T., and H.G. Perros "Approximate Analysis of Arbitrary Configurations of Queueing Networks with Blocking" Annals of Oper. Res. 9 (1987) 481-509.

Altiok, T., and S.S. Stidham "A note on Transfer Line with Unreliable Machines, Random Processing Times, and Finite Buffers" IIE Trans., Vol. 14, 4 (1982) 125-127.

Altiok, T., and S.S. Stidham "The Allocation of Interstage Buffer Capacities in Production Lines" IIE Trans., Vol. 15 (1983) 292-299.

Ammar, M.H., and S.B. Gershwin "Equivalence Relations in Queueing Models of Assembly/Disassembly Networks" Research Report, Georgia Institute of Technology.

Ammar, M.H., and S.B. Gershwin "Equivalence Relations in Queueing Models of Fork/Join Networks with Blocking" Performance Evaluation, Vol. 10 (1989) 233-245.

Balsamo, S. "Decomposability for General Markovian Networks", in *Mathematical Computer Performance and Reliability*, (G. Iazeolla, P.J. Courtois, A. Hordijk Eds.), North Holland, 1984.

Balsamo, S. "Properties and analysis of queueing network models with finite capacities", in *Performance Evaluation of Computer and Communication Systems* (L. Donatiello, R. Nelson Eds.) Lecture Notes in Computer Science, 729, 1994, Springer-Verlag.

Balsamo, S., and C. Clò "State distribution at arrival times for closed queueing networks with blocking" Technical Report TR-35/92, Dept. of Comp. Sci., University of Pisa, 1992.

Balsamo, S., C. Clò, and L. Donatiello "Cycle Time Distribution of Cyclic Queueing Network with Blocking", in *Queueing Networks with Finite Capacities* (R.O. Onvural and I.F. Akyildiz Eds.), Elsevier, 1993, and Performance Evaluation, Vol. 17 (1993) 159-168.

Balsamo, S., and C. Clò "Delay distribution in a central server model with blocking", Technical Report TR-14/93, Dept. of Comp. Sci., University of Pisa, 1993.

Balsamo, S., C. Clò "A Convolution Algorithm for Product Form Queueing Networks with Blocking" Annals of Operations Research, Vol. 79 (1998) 97-117.

Balsamo, S., V. De Nitto, G. Iazeolla "Identity and Reducibility Properties of Some Blocking and Non-Blocking Mechanisms in Congested Networks" in *Flow Control of Congested Networks*, (A.R. Odoni, L. Bianco, G. Szego Eds.), NATO ASI Series, Comp. and System Science, Vol.F38, Springer-Verlag, 1987.

Balsamo, S., and V. De Nitto Personè "Closed queueing networks with finite capacities: blocking types, product-form solution and performance indices" Performance Evaluation, Vol. 12, 4 (1991) 85-102.

Balsamo, S. and V. De Nitto Personè "A survey of Product-form Queueing Networks with Blocking and their Equivalences" Annals of Operations Research, vol. 48 (1994) 31-61.

Balsamo, S., and L. Donatiello "On the Cycle Time Distribution in a Two-stage Queueing Network with Blocking" IEEE Transactions on Software Engineering, Vol. 13 (1989) 1206-1216.

Balsamo, S., and L. Donatiello "Two-stage Queueing Networks with Blocking: Cycle Time Distribution and Equivalence Properties", in *Modelling Techniques and Tools for Computer Performance Evaluation* (R. Puigjaner, D. Potier Eds.) Plenum Press, 1989.

Balsamo, S., L. Donatiello and N. Van Dijk "Bounded performance analysis of parallel processing systems" IEEE Trans. on Par. and Distr. Systems, Vol. 9 (1998) 1041-1056.

Balsamo, S., and G. Iazeolla "Some Equivalence Properties for Queueing Networks with and without Blocking" in *Performance '83* (A.K. Agrawala, S.K. Tripathi Eds.) North Holland, 1983.

Balsamo, S., and G. Iazeolla "An extension of Norton's Theorem for Queueing Networks" IEEE Transactions on Software Engineering, Vol.8 (1982) 298-305.

Balsamo, S., and G. Iazeolla "Synthesis of Queueing Networks with Block and State-dependent Routing" Computer System Science and Engineering, Vol. 1 (1986).

Balsamo, S., and B. Pandolfi "Bounded Aggregation in Markovian Networks" in *Computer Performance and Reliability* (G. Iazeolla, P.J. Courtois, O. Boxma Eds.), North Holland, 1988.

Balsamo, S., and A. Rainero "Approximate Performance Analysis of Queueing Networks with Blocking: A Comparison" UDMI/05/98/RR, Dept. Math and Comp. Sci., University of Udine, March 1998.

Baskett, F., K.M. Chandy, R.R. Muntz, and G. Palacios "Open, closed, and mixed networks of queues with different classes of customers" J. of ACM, 22 (1975) 248-260.

Boucherie, R. "Norton's Equivalent for queueing networks comprised of quasireversible components linked by state-dependent routing" Performance Evaluation, Vol. 32 (1998) 83-99.

Boucherie, R., and N.M. Van Dijk "A generalization of Norton's theorem for queueing networks" Queueing Systems, Vol. 13 (1993) 251-289.

Boucherie, R., and N. Van Dijk "On the arrival theorem for product form queueing networks with blocking" Performance Evaluation, 29 (1997) 155-176.

Bouchouch, A., Y. Frein and Y. Dallery "Performance evaluation of closed tandem queueing networks with finite buffers" Performance Evaluation, Vol. 26 (1996) 115-132.

Boxma, O., and A.G. Konheim "Approximate analysis of exponential queueing systems with blocking" Acta Informatica, 15 (1981) 19-66.

Boxma, O., and H. Daduna "Sojourn time distribution in queueing networks" in 'Stochastic Analysis of computer and Communication Systems' (H. Takagi Ed.) North Holland, 1990.

Brandwajn, A., and Y.L. Jow "An approximation method for tandem queueing systems with blocking" Operations Research, Vol. 1 (1988) 73-83.

Buzacott, J.A. "Automatic Transfer Lines with Buffer Stocks" Int. J. Prod. Res., Vol. 5 (1967) 183-200.

Buzacott, J.A. "Prediction of the Efficiency of Production Systems without Internal Storage" Int. J. Prod. Res., Vol. 6 (1968) 173-188.

Buzacott, J.A. "Queueing Models of Kanban and MRP controlled production systems" Engineering Costs and Production Economics, Elsevier Science Pub., Vol. 17 (1989) 3-20.

Buzacott, J..A. "The Role of Inventory Banks in Flow Line Production Systems" Int. J. Prod. Res., Vol. 9 (1971) 425-436.

Buzacott, J..A., and J.G. Shanthikumar "Design of Manufacturing Systems using Queueing Models" Queueing Systems: Theory and Applications, (1992).

Buzen, J.P. "Computational Algorithms for Closed Queueing Networks with exponential servers" Comm. ACM, Vol. 16 (1973) 527-531.

Caseau, P., and G. Pujolle "Throughput capacity of a sequence of transfer lines with blocking due to finite waiting room" IEEE Trans. on Softw. Eng. 5 (1979) 631-642.

Chandy, K.M., U. Herzog, and L. Woo "Parametric analysis of queueing networks" IBM J. Res. Dev., 1 (1975) 36-42.

Chandy, K.M., J.H. Howard, and D. Towsley "Product form and local balance in queueing networks" J. ACM, Vol. 24 (1977) 250-263.

Chandy, K.M., and A.J. Martin "A characterization of product-form queueing networks" J. ACM, Vol.30 (1983) 286-299.

Chandy, K.M., and C.H. Sauer "Approximate Methods for Analyzing Queueing Network Models of Computing Systems" ACM Computing Surveys Vol. 10 (1978) 281-317.

Cheng, D.W. "Analysis of a tandem queue with state dependent general blocking: a GSMP perspective" Performance Evaluation, Vol. 17 (1993) 169-173.

Christofides, N. *Graph Theory - An Algorithmic Approach.* Academic Press, London, 1975.

Choukri, T. "Exact Analysis of Multiple Job Classes and Different Types of Blocking" in *Queueing Networks with Finite Capacities* (R.O. Onvural and I.F. Akyidiz Eds.),Elsevier, 1993.

Clò, C. "MVA for Product-Form Cyclic Queueing Networks with RS Blocking" Annals of Operations Research, Vol. 79 (1998).

Coffman, E.G., M.J. Elphick, and A. Shoshani "System Deadlocks" ACM Computing Surveys, Vol. 2 (1971) 67-78.

Cohen, J.W *The Single Server Queue*, North Holland Publishing Company, Amsterdam, 1969.

Courtois, P.J. *Decomposability: Queueing and Computer System Applications*, Academic Press, Inc, New York, 1977.

Courtois, P.J., and P. Semal "Computable bounds for conditional steady-state probabilities in large Markov chains and queueing models" IEEE Journal on SAC, Vol. 4 (1986) 920-936.

Cox, D.R. "A Use of Complex Probabilities in the Theory of Stochastic Processes" in Proc. Cambridge Philosophical Society, 51 (1955) 313-319.

Daduna, H. "Busy Periods for Subnetworks in Stochastic Networks: Mean Value Analysis" J. ACM, Vol. 35 (1988) 668-674.

Dallery, Y., and Y. Frein "A decomposition method for the approximate analysis of closed queueing networks with blocking", Proc. First Int. Workshop on Queueing Networks with Blocking, (H.G. Perros and T. Altiok Eds.) North Holland (1989).

Dallery, Y., and Y. Frein "On decomposition methods for tandem queueing networks with blocking" Operations Research, Vol. 14 (1993) 386-399.

Dallery, Y., and S.B. Gershwin "Manufacturing flow line systems: A review of models and analytical results" Queueing Systems, Vol. 12 (1992) 3-94.

Dallery, Y., Z. Liu, and D.F. Towsley "Properties of fork/join queueing networks with blocking under various blocking mechanisms" Techn, Report, MASI.92, Université Pierre et Marie Curie, France, April, 1992.

Dallery, Y., Z. Liu, and D.F. Towsley "Equivalence, reversibility, symmetry and concavity properties in fork/join queueing networks with blocking" Techn, Report, MASI.90.32, Université Pierre et Marie Curie, France, June, 1990 and J. of the ACM, Vol. 41 (1994) 903-942.

Dallery, Y., and D.F. Towsley "Symmetry property of the throughput in closed tandem queueing networks with finite buffers" Op. Res. Letters, Vol. 10 (1991) 541-547.

Dallery, Y., and D.D. Yao "Modelling a system of flexible manufacturing cells" in: Modeling and Design of Flexible Manufacturing Systems (Kusiak Ed.) North-Holland (1986) 289-300.

De Nitto Personè, V. "Topology related index for performance comparison of blocking symmetrical networks" European J. of Oper. Res., Vol. 78 (1994) 413-425.

De Nitto Personè, V., and D. Grillo "Managing Blocking in Finite Capacity Symmetrical Ring Networks" in Proc. 3rd Conference on Data and Communication Systems and Their Performance, Rio de Jenerio, Brasil, 1987.

De Nitto Personè, V., and V. Grassi "An analytical model for a parallel fault-tolerant computing system" Performance Evaluation, Vol. 38 (1999) 201-218.

Diehl, G.W. "A Buffer Equivalency Decomposition Approach to Finite Buffer Queueing Networks" Ph.D. thesis, Eng. Sci., Harvard University, 1984.

Frein, Y., and Y. Dallery "Analysis of Cyclic Queueing Networks with Finite Buffers and Blocking Before Service", *Performance Evaluation*, Vol. 10 (1989) 197-210.

Gavish, B. "A General Model for Topological Design of Computer Networks" in Proc. Globecom'86, 1986.

Gavish, B., and I. Neuman "Capacity and Flow Assignments in Large Computer Networks" in Proc. IEEE-Infocom'86, 1986, 275-284.

Gelenbe, E., and I. Mitrani. *Analysis and Synthesis of Computer Systems.* Academic Press, Inc., London, 1980.

Gershwin, S. B. "An Efficient Decomposition Method for the Approximate Evaluation of Production Lines with Finite Storage Space" in Proc. Sixth INRIA Conf. Proc., Nice, France, 1984.

Gershwin, S. B. "An efficient decomposition method for the approximate evaluation of tandem queues with finite storage space and blocking" Oper. Res., Vol. 35 (1987) 291-305.

Gershwin, S. B., and U. Berman "Analysis of Transfer Lines Consisting of Two Unreliable Machines with Random Processing Times and Finite Storage Buffers" AIIE Trans., 1Vol. 3- (1981) 2-11.

Grillo, D., and V. De Nitto Personè "Impact of blocking policies in closed, exponential and symmetrical ring networks" Tech. Report 4B0586, Fondazione Ugo Bordoni, Roma, 1986.

Gordon, W.J., and G.F. Newell "Cyclic queueing systems with restricted queues" Oper. Res., Vol. 15 (1967) 286-302.

Gordon, W.J., and G.F. Newell "Cyclic queueing systems with Exponential Servers" Oper. Res., Vol. 15 (1967) 254-265.

Gross, D., and C.M. Harris *Fundamentals of Queueing Theory*. John Wiley and Sons, Inc., New York, 1974.

Gün, L., and A.M. Makowski "An approximation method for general tandem queueing systems subject to blocking" Proc. First Int. Workshop on Queueing Networks with Blocking, (H.G. Perros and T. Altiok Eds.) North Holland, 1989,147-171.

Heidelberger, P., and S.K. Trivedi "Queueing Network models for parallel processing with synchronous tasks" IEEE Trans. on Comp., Vol. 31 (1982) 1099-1109.

Hillier, F.S., and W. Boling "The Effect of Some Design Factors on the Efficiency of Production Lines with Variable Operation Times" J. Ind. Eng., Vol. 7 (1966) 651-658.

Hillier, F.S., and W. Boling "Finite queues in series with exponential or Erlang service times - a numerical approach" Oper. Res., Vol. 15 (1967) 286-303.

Hillier, F.S., and W. Boling "On the Optimal Allocation of Work in Symmetrically Unbalanced Production Line Systems with Variable Operation Times" Mgmt. Sci., Vol. 25 (1979) 721-728.

Hillier, F.S., and K.C. So "The assignment of extra servers to stations in tandem queueing systems with small or no buffers" Performance Evaluation, Vol. 10 (1989) 213-231.

Hillier, F.S., and K.C. So "On the Optimal Allocation of Work in Symmetrically Unbalanced Production Line Systems with Variable Operation Times" Mgmt. Sci., Vol. 25 (1979) 721-728.

Highleyman, W. H. *Performance Analysis of Transaction Processing Systems*. Prentice Hall, Inc., Englewood Cliffs, New Jersey, 1989.

Hordijk, A., and N. Van Dijk "Networks of queues with blocking", in: Performance '81 (K.J. Kylstra Ed.) North Holland (1981) 51-65.

Hordijk, A., and N. Van Dijk "Adjoint Processes, Job Local Balance and Insensitivity of Stochastic Networks", Bull:44 session, Int. Stat. Inst., Vol. 50 (1982) 776-788.

Hordijk, A. , and N. Van Dijk "Networks of Queues: Part I-Job Local Balance and the Adjoint Process", "Part II-General Routing and Service Characteristics" in Proc. Int. Conf. Modeling Comput. Sys., 60 (1983) 158-205.

Jackson, J.R. "Jobshop-like Queueing Systems" Mgmt. Sci., Vol. 10 (1963) 131-142.

Jafari, M. A. and J.G. Shanthikumar "Determination of Optimal Buffer Storage Capacities and Optimal Allocation in Multistage Automatic Transfer Lines" IIE Trans., Vol. 21 (1989) 130-135.

Jennings, A. *Matrix Computation for Engineers and Scientists*. John Wiley and Sons, Inc., New York, 1977.

Jun, K.P. "Approximate Analysis of Open Queueing Networks with Blocking", Ph.D. thesis, Operations Research Program, North Carolina State University, 1988.

Kant, K. *Introduction to Computer System Performance Evaluation*. McGraw-Hill, 1992.

Kelly, K.P. *Reversibility and Stochastic Networks*. John Wiley and Sons Ltd., Chichester, England, 1979.

Kendall, D. G. "Stochastic Processes Occurring in the Theory of Queues and Their Analysis by the Method of Imbedded Markov Chains" Ann. Math. Statist., 24 (1953) 338-354

Kingman, J.F.C. "Markovian population process" J. Appl. Prob., Vol. 6 (1969) 1-18.

Kleinrock, L. *Queueing Systems-Vol II*, John Wiley and Sons, New York, 1976.

Kritzinger, P., S., Van Wyk, and A. Krzesinski "A generalization of Norton's theorem for multiclass queueing networks" Performance Evaluation, Vol. 2 (1982) 98-107.

Krzesinski, A.E. "Multiclass queueing networks with state-dependent routing" Performance Evaluation, Vol.7 (1987) 125-145.

Konhein, A.G., and M. Reiser "A queueing model with finite waiting room and blocking" SIAM J. of Comput, Vol. 7 (1978) 210-229.

Kouvatsos, D.D. "Maximum Entropy Methods for General Queueing Networks" in Proc. Modeling Tech. and Tools for Perf. Analysis, (Potier Ed.), North Holland, 1983, 589-608.

Kouvatsos, D., and I.U. Awan "Arbitrary closed queueing networks with blocking and multiple job classes" Proc. Third International Workshop on Queueing Networks with Finite Capacity, Bradford, UK, 6-7 July, 1995.

Kouvatsos, D., and S.G. Denazis "Entropy maximized queueing networks with blocking and multiple job classes" Performance Evaluation, Vol. 17 (1993) 189-205.

Kouvatsos, D.D., and N.P. Xenios "Maximum Entropy Analysis of General Queueing Networks with Blocking", in First International Workshop on Queueing Networks with Blocking, (Perros and Altiok Eds), Elsevier Science Publishers North Holland, 1989.

Kubat, P., and U. Sumita "Buffers and Backup Machines in Automatic Transfer Lines" Int. J. Prod. Res., Vol. 23 (1985) 1259-1280.

Kundu, S., and I. Akyildiz "Deadlock free buffer allocation in closed queueing networks" Queueing Systems Journal, Vol. 4 (1989) 47-56.

Jun, K.P., and H.G. Perros "An approximate analysis of open tandem queueing networks with blocking and general service times" Europ. Journal of Operations Research, Vol. 46 (1990) 123-135.

Lam, S.S. "Queueing networks with capacity constraints" IBM J. Res. Develop., Vol. 21 (1977) 370-378.

Lavenberg, S.S. *Computer Performance Modeling Handbook*. Prentice Hall, 1983.

Lavenberg, S.S., and M. Reiser "Stationary State Probabilities at Arrival Instants for Closed Queueing Networks with multiple Types of Customers" *J. Appl. Prob.*, Vol. 17 (1980) 1048-1061.

Law, A.M., and W.D. Kelton *Simulation Modeling and Analysis*. Mc Graw Hill, New York, 1982.

Lee, H.S., and S. M. Pollock "Approximation analysis of open acyclic exponential queueing networks with blocking" Operations Research, Vol. 38 (1990) 1123-1134.

Lee, H.S., A. Bouhchouch, Y. Dallery and Y. Frein "Performance Evaluation of open queueing networks with arbitrary configurations and finite buffers" Proc. Third International Workshop on Queueing Networks with Finite Capacity, Bradford, UK, 6-7 July, 1995.

Little, J.D.C. "A Proof of the Queueing Formula L=λW" Oper. Res., Vol. 9 (1961) 383-387

Liu, X.G., and J.A. Buzacott "A balanced local flow technique for queueing networks with blocking" Proc. First Int. Workshop on Queueing Networks with Blocking (H. Perros and T. Altiok Eds.), North Holland, 1989, 87-104.

Liu, X.G., and J.A. Buzacott "A decomposition related throughput property of tandem queueing networks with blocking" Queueing Systems, Vol. 13 (1993) 361-383.

Liu, X.G., L. Zwang and J.A. Buzacott "A decomposition method for throughput analysis of cyclic queues with production blocking" in *Queueing Networks with Finite Capacity* (R. O. Onvural and I.F. Akyildiz Eds.) North Holland, 1993, 253-266.

MacGregor Smith, J., and S. Daskalaki "Buffer Space Allocation in Automated Assembly Lines" Operations Research, Vol. 36 (1988) 343-358.

Marie, R. "An Approximate Analytical Method for General Queueing Networks" IEEE Trans. on Software Engineering, Vol. 5 (1979) 530-538.

Melamed, B. "A note on the reversibility and duality of some tandem blocking queueing systems" Manage. Sci., Vol. 32 (1986) 1648-1650.

Minoura, T. "Deadlock Avoidance Revisited" Journal of ACM, Vol. 29 (1982) 1023-1048.

Mishra, S., and S.C. Fang "A maximum entropy optimization approach to tandem queues with generalized blocking" Performance Evaluation, Vol. 30 (1997) 217-241.

Mitra, D. , and I. Mitrani "Analysis of a Novel Discipline for Cell Coordination in Production Lines" AT&T Bell Labs Res. Rep., 1988.

Mitra, D., and I. Mitrani " Analysis of a Kanban discipline for cell coordination in production lines I" Management Science, Vol. 36 (1990) 1548-1566.

Mitra, D., and I. Mitrani "Analysis of a Kanban discipline for cell coordination in production lines II: Stochastic demands" Operations Research, Vol. 36 (1992) 807-823.

Muntz, R.R. "Queueing Networks: A Critique of the State of the Art Directions for the Future" ACM Comp. Surveys, Vol. 10 (1978) 353-359.

Muth, E.J. "The reversibility property of production lines" Manage. Sci., Vol. 25 (1979) 152-158.

Nelson, R., D. Towsley and A. Tantawi "Performance analysis of parallel processing systems" IEEE Trans. on Softw. Eng., Vol. 14 (1988) 532-540.

Neuts, M.F. "Two queues in series with a finite intermediate waiting room" J. Appl. Prob., 5 (1986) 123-142.

Onvural, R.O. "Closed Queueing Networks with Finite Buffers" Ph.D. thesis, CSE/OR, North Carolina State University, 1987.

Onvural, R.O. "On the Exact Decomposition of Exponential Closed Queueing Networks with Blocking" in First International Workshop on Queueing Networks with Blocking, (Perros and Altiok Eds), Elsevier Science Publishers, North Holland, 1989.

Onvural, R.O. "Some Product Form Solutions of Multi-Class Queueing Networks with Blocking' Performance Evaluation, Special Issue on Queueing Networks with Blocking, (Akyildiz and Perros Eds), 1989.

Onvural, R.O. "A Note on the Product Form Solutions of Multiclass Closed Queueing Networks with Blocking" Performance Evaluation, Vol.10 (1989) 247-253.

Onvural, R.O. "Survey of Closed Queueing Networks with Blocking" ACM Computing Surveys, Vol. 22, 2 (1990) 83-121.

Onvural, R.O. Special Issue on Queueing Networks with Finite Capacity, Performance Evaluation, Vol. 17, 3 (1993).

Onvural, R.O., and H.G. Perros "On Equivalencies of Blocking Mechanisms in Queueing Networks with Blocking" Oper. Res. Letters, Vol. 5 (1986) 293-298.

Onvural, R.O., and H.G. Perros "Equivalencies Between Open and Closed Queueing Networks with Finite Buffers" in Proc. International Seminar on the Performance Evaluation of Distributed and Parallel Systems, Kyoto, Japan. Also to appear in Performance Evaluation, 1988.

Onvural, R.O., and H.G. Perros "Some equivalencies on closed exponential queueing networks with blocking" Performance Evaluation, Vol.9 (1989) 111-118.

Onvural, R.O., and H.G. Perros "Throughput Analysis in Cyclic Queueing Networks with Blocking" IEEE Trans. Software Engineering, Vol. 15 (1989) 800-808.

Papadopoulus, H.T., and C. Heavey "Queueing Theory in manufacturing systems analysis and design: A classification of models for production and transfer lines" Europ. J. of Oper. Res., Vol. 92 (1996) 1-27.

Perros, H.G. "Queueing Networks with Blocking: A Bibliography" ACM Sigmetrics, Performance Evaluation Review, Vol. 12 (1984) 8-12.

Perros, H.G. "Open Queueing Networks with Blocking" in Stochastic Analysis of Computer and Communications Systems, (Takagi Ed.), Elsevier Science Publishers, North Holland, 1989.

Perros, H.G. Queueing networks with blocking. Oxford University Press, 1994.

Perros, H.G., and T. Altiok "Approximate analysis of open networks of queues with blocking: tandem configurations" IEEE Trans. on Software Eng., Vol. 12 (1986) 450-461.

Perros, H.G., A. Nilsson, and Y.C. Liu "Approximate Analysis of Product Form Type Queueing Networks with Blocking and Deadlock" Performance Evaluation (1989).

Perros, H.G., and P.M. Snyder "A computationally efficient approximation algorithm for analyzing queueing networks with blocking" Performance Evaluation, Vol. 9 (1988/89) 217-224.

Pittel, B. "Closed Exponential Networks of Queues with Saturation: The Jackson Type Stationary Distribution and Its Asymptotic Analysis" Math. Oper. Res., Vol. 4 (1979) 367-378.

Raghavendra, C.S., and J.A. Silvester "A Survey of multi-connected loop topologies for local computer networks" Computer Networks and ISDN Systems, Vol. 11 (1986) 29-42.

Ree, D.A., and H.D. Shwetman "Cost-performance bounds for multicomputer networks" IEEE Trans. on Computer, Vol. 32 (1983) 83-95.

Reed, D.A., and H.D. Shwetman "Cost-performance bounds for multicomputer networks" IEEE Trans. on Computer, Vol. 32 (1983) 83-95.

Reed D.A., and D.C. Grunwald "The performance of Multicomputer Interconnection Networks" Computer, (1987) 63-73.

Reiser, M. "A Queueing Network Analysis of Computer Communications Networks with Window Flow Control" IEEE Trans. on Comm., Vol. 27 (1979) 1199-1209.

Sereno, M. "Mean Value Analysis of product form solution queueing networks with repetitive service blocking" Performance Evaluation, Vol. 36-37 (1999) 19-33.

Sevastyonov, B.A. "Influence of Storage Bin Capacity on the Average, Standstill Time of Production Line" Theory of Probability and Its Applications, Vol. 7 (1962) 429-438.

Sevcik, K.S., and I. Mitrani "The Distribution of Queueing Network States at Input and Output Instants" J. of ACM, Vol. 28 (1981) 358-371.

Shanthikumar, G.J., and D.D. Yao "Monotonicity Properties in Cyclic Queueing Networks with Finite Buffers" in First International Workshop on Queueing Networks with Blocking, (Perros and Altiok Eds), Elsevier Science Publishers, North Holland, 1989.

Sheskin, T. J. "Allocation of Interstage Storage Along an Automated Production Line" AIIE Trans., Vol. 8 (1976) 146-152.

Solomon, S.L. *Simulation of Waiting Line Systems*. Prentice Hall, Englewood Cliffs, New Jersey, 1983.

Soyster, A.L., J.W. Schmidt, and M.W. Rohrer "Allocation of Buffer Capacities for a Class of Fixed Cycle Production Lines" AIIE Trans., Vol. 11 (1979) 140-146.

Suri, R., and G.W. Diehl "A New Building Block for Performance Evaluation of Queueing Networks with Finite Buffers" in Proc. ACM Sigmetrics on Measurement and Modeling of Computer Systems (1984) 134-142.

Suri, R., and G.W. Diehl "A Variable Buffer Size Model and Its Use in Analytical Closed Queueing Networks with Blocking" Management Science, Vol. 32 (1986) 206-225.

Towsley, D.F. "Queueing network models with state-dependent routing" J. ACM, Vol. 27 (1980) 323-337.

Trivedi, K.S. *Probability and Statistics with Reliability, Queueing and Computer Science Applications*. Prentice Hall, Englewood Cliffs, New Jersey, 1982.

Van Dijk, N. "On 'stop = repeat' servicing for non-exponential queueing networks with blocking" J. Appl. Prob., Vol. 28 (1991) 159-173.

Van Dijk, N. "'Stop = recirculate' for exponential product form queueing networks with departure blocking" Oper. Res. Lett., Vol. 10 (1991) 343-351.

Van Dijk, N. "On the Arrival Theorem for communication networks" Computer Networks and ISDN Systems, Vol. 25 (1993) 1135-1142.

Van Dijk, N.M., and H.C. Tijms "Insensitivity in Two Node Blocking Models with Applications" in *Teletraffic Analysis and Computer Performance Evaluation* (Boxma, Cohen and Tijms Eds.), Elsevier Science Publishers, North Holland, 1986, 329-340.

Vantilborgh, H., "Exact aggregation in exponential queueing networks" Journal of the ACM, Vol. 25 (1978) 620-629.

Whitt, W. "Open and Closed Models for Networks of Queues" AT&T Bell Labs Tec. J., Vol. 63 (1984) 1911-1979.

Whittle, P. "Partial balance and insensitivity" J. Appl. Prob. 22 (1985) 168-175.

Yamashita, H., and S. Suzuki "An Approximate Solution Method for Optimal Buffer Allocation in Serial n-stage Automatic Production Lines" Trans. Japan. Soc. Mech. Eng., Vol. 53-C (1987) 807-814, (in Japanese).

Yamazaki, G., and H. Sakasegawa "Problems of duality in tandem queueing networks" Ann. Inst. Math. Stat., Vol. 27 (1975) 201-212.

Yao, D.D., and J.A. Buzacott "Modeling a Class of State Dependent Routing in Flexible Manufacturing Systems" Annals of Operations Research, Vol. 3 (1985) 153-167.

Yao, D.D., and J.A. Buzacott "Queueing Models for Flexible Machining Station Part I: Diffusion Approximation" Eur. J. Operations Research, Vol. 19 (1985) 233-240.

Yao, D.D., and J.A. Buzacott "Queueing Models for Flexible Machining Stations Part II: The Method of Coxian Phases" Eur. J. Operations Research, Vol. 19 (1985) 241-252.

Yao, D.D., and J.A. Buzacott "The Exponentialization Approach to Flexible Manufacturing System Models with General Processing Times" Eur. J. of Operations Research, Vol. 24 (1986) 410-416.

TABLE OF SYMBOLS

Symbol	Definition	pages
$a(n)$	arrival function	25
a_i	minimum population in node i	64
$a_r(\mathbf{N})$	arrival function for multichain networks	36
$b_i(n_i)$	blocking function, probability that a job arriving at node i is accepted when there are n_i customers in it	27
B_i	node i capacity	27
B_W	maximum population allowed in a subnetwork W	27
C	number of classes	36
$d(n)$	departure blocking function	33
$\delta(n_i)$	indicator function for empty condition	66
E	state space of queueing system	7
E_r	class set of chain r for a multiclass network	36
$E_{RS\text{-}RD}$	state space of the Markov process of the network with RS-RD blocking	66
$440E_{BAS}$	state space of the Markov process of the network with BAS blocking	69
$E_{BBS\text{-}SO}$	state space of the Markov process of the network with BBS-SO blocking	74
$E_{BBS\text{-}SNO}$	state space of the Markov process of the network with BBS-SNO blocking	77
$E_{BBS\text{-}O}$	state space of the Markov process of the network with BBS-O blocking	79
E_{Stop}	state space of the Markov process of the network with Stop blocking	84
$E_{Recirculate}$	state space of the Markov process of the network with Recirculate blocking	84
E_{Het}	state space of the Markov process of a heterogeneous network	87
$f_i(n_i, K_i)$	arbitrary non-negative function for load dependent service	26
$F_i(t)$	service time distribution at node i	26
$F_{is}(t)$	service time distribution at node i and class s	37
K_i	number of identical servers at node i	26
L_i	mean queue length of node i	38
λ	average arrival rate	5
M	number of nodes in the network	9

m_i	list of nodes blocked by node i according to BAS mechanism	68
μ_i	service rate at node i	6
μ_{is}	service rate at node i and class s	37
n_i	number of jobs at node i	10
n_{ir}	number of jobs in node i and chain r	37
N	total number of customers in the network	9
$NS_{i,k}$	number of node i servers that are servicing jobs destined to node k, according to BBS mechanism	73
N_r	total number of chain r customers in the network	36
$N=(N_1,É,N_R)$	population vector for multichain networks	36
$P=[p_{ij}]$	matrix of routing probabilities	14
$P=[p_{(i,s)(j,t)}]$	matrix of routing probabilities for multichain networks, $1^2s,t^2C$	36
$^1(S)$	steady state queue length distribution	11
$^1_i(n)$	steady state marginal queue length probability	13
Q	process rate matrix	7
R	number of routing chains	9
S	network state	10
S_i	node i state	10
s_i	number of servers of node i blocked by a full destination node, according to BAS notation	68
T	mean residence time in the network	43
T_i	node i mean response time	14
t_i	customer passage time through the node i	38
$T(i,j)$	mean passage time from node i to node j	43
U_i	node i utilization	38
U_i^e	node i effective utilization	39
X_i	node i throughput	38
X_i^e	node i effective throughput	41
x_i	visit ratio, the *relative* arrival rate of customers to node i	14
x_{ir}	visit ratio, the *relative* arrival rate of chain r customers to node i	11
$\xi_i(n_i)$	stationary queue length distribution of node i at arrival time	38
z_i	number of active servers at node i	38
$\zeta_i(n_i,z_i)$	stationary joint distribution of variables n_i and z_i	38

INDEX